Havi... ...c.la.... ... ...Science Communication from Birkbeck Colle... ...niversity of London – as well as a PhD in Immunology – Catherine Whitlock now works as a freelance writer.

Her most recent work, including her latest book, can be found at www.catherinewhitlock.co.uk. She is a Chartered Biologist and a member of the British Society for Immunology and the Association of British Science Writers. She lives in Kent with her husband and three children.

Dr Rhodri Evans studied Physics at Imperial College London before gaining his PhD in Astrophysics from Cardiff University. He has taught at a number of universities around the world and is the author of numerous popular-science articles. He speaks at conferences and is a regular contributor to the BBC on Physics and Astronomy. His popular blog can be found at thecuriousastronomer.wordpress.com

Ten Women
Who Changed Science,
and the World

....

CATHERINE WHITLOCK

AND

RHODRI EVANS

Foreword by Athene Donald, Professor of Experimental
Physics, University of Cambridge, and Master of
Churchill College

ROBINSON

ROBINSON

First published in Great Britain in 2019 by Robinson

1 3 5 7 9 10 8 6 4 2

Copyright © Catherine Whitlock and Rhodri Evans, 2019
Illustrations by Stephen Dew, 2019

The moral rights of the authors have been asserted.

A CIP catalogue record for this book
is available from the British Library.

ISBN: 978-1-47213-743-2

Typeset in Scala byHewer Text UK Ltd, Edinburgh
Printed and bound in Great Britain by Clays Ltd, Elcograf S.p.A

Papers used by Robinson are from well-managed
forests and other responsible sources.

Robinson
An imprint of
Little, Brown Book Group
Carmelite House
50 Victoria Embankment
London EC4Y 0DZ

An Hachette UK Company
www.hachette.co.uk

www.littlebrown.co.uk

Contents

Foreword

It is over a century since the first woman received a Nobel Prize in science. In that time, since 1911 when Marie Curie received that accolade, only a further eighteen women have been likewise so honoured (including Marie Curie who won it twice) and only a single woman in the UK has been so honoured. When she was – Dorothy Hodgkin in 1964 – did the press regard her in the same light as they would a man in the same position? Absolutely not. The *Daily Telegraph* announced 'British woman wins Nobel Prize – £18,750 prize to mother of three'. The *Daily Mail* was even briefer in its headline 'Oxford housewife wins Nobel'. The *Observer* commented in its write-up 'affable-looking housewife Mrs Hodgkin' had won the prize 'for a thoroughly unhousewifely skill: the structure of crystals of great chemical interest'. It makes for depressing reading fifty years later, but we have nothing more up to date to leaven the message. Two more women winning prizes in 2018 is a step in the right direction, but hardly proof that the situation is transformed.

Dorothy Hodgkin, featured as one of the ten outstanding women who have contributed so much to the world of science in this book, would not have had time to consider whether or not she was a feminist (although in later life she was very visibly a pacifist). She only wanted to get on with the job of what really interested her: the structure of biological molecules. As she put it, she just wanted to 'live simply and do serious things', and serious things she most certainly did, solving the three-dimensional structures of insulin, vitamin B12 and penicillin amongst other complex molecules. As a woman working in a man's world, she simply dedicated herself to achieving as much as she could and small matters like pregnancy were not allowed to get in the way. When married, but still working under her maiden name of Crowfoot, she presented a key paper at a major meeting at the Royal Society in 1938 when eight months pregnant. A long-term collaborator, (and another Nobel Prize winner) Max Perutz, referred to her appearance at this meeting in his speech at her memorial service: 'Dorothy lectured in that state as if it were the most natural thing in the world, without any pretence of trying to be unconventional, which it certainly was at the time.'

In this book, her life and those of nine other remarkable women, including Marie Curie, are explored – women from around the world and from very different cultures and backgrounds. It is interesting to see what common features their lives share and what that might mean for young women growing up now. At the top level of sciences, particularly the physical sciences, there is still a dire paucity of women. Diversity – amongst Nobel Prize winners in particular, but also amongst the movers and shakers (and winners) in science – is still rather limited. The women chosen for this book are all dead, not living role models who might be seen on TV or interviewed in the press (let alone liked on social media): the authors felt that that distance provides perspective and understanding.

In the days before superfast global communication, these women's science and the impact they made often remained hidden and to a certain extent unrecognised, sometimes by their peers and almost invariably by the general public. Even in today's era of highly accessible information, their importance and impact are not well known. They and their work should be better appreciated because they were ground-breaking trailblazers, whether or not they would have recognized that at the time.

Luck plays a role in every scientist's life, whether or not they are prepared to admit it. In the case of Gertrude Elion it was her father's suggestion that – after repeatedly failing to get even as far as job interviews after obtaining her MSc in Chemistry – she ring up Burroughs Wellcome, simply because he was familiar with the company because they made a painkiller he used in his dental practice. There, Elion quickly found her feet and stayed for many years researching novel 'designer drugs', for which she ultimately won the Nobel Prize in 1988. The Chinese-American physicist Chien-Shiung Wu said of her own work that 'Relying purely on ingenuity, determination and luck, three of us (an enthusiastic chemist, a dedicated student and myself) worked together uninterruptedly to grow about ten large perfect translucent CMN single crystals by the end of three weeks.' Growing crystals of complex molecules is something of a black art, which is why luck enters into it. But she was also unlucky in that the Nobel Committee overlooked her strong credentials; she enters the group of women – to which Lise Meitner from this book also belongs – who are so often identified as those who did *not* win a Nobel Prize when they were more than deserving. Here, lack of luck in the lottery of winning big prizes also enters into the dimension, along with the bias often attributed to the Swedish committee.

The determination Chien-Shiung Wu mentions is also a crucial character trait absolutely required for success for anyone in science. Determination is often epitomised by the hard work that Marie Curie, with her husband Pierre, put in to extract from the bulk mineral pitchblende the trace components of the elements ultimately known as thorium and polonium that give rise to high levels of radioactivity, levels she recognised as being much higher than that due to uranium alone. Marie Curie is of course *the* female scientist that most members of the public are likely to be able to name and the only woman accorded the honour of two Nobel Prizes, one in Physics and one in Chemistry.

Rachel Carson was another woman who exhibited enormous determination to bring her concerns about environmental pollution to the wider public, however much this steeliness may have been hidden behind a quiet exterior. Without this strength of mind coupled with beautiful writing skills, it is unlikely that the dangers associated with pesticides such as DDT would have been recognized – and acted upon – anything like as fast.

Above all else, though, scientists must harness their creativity and imagination. Research and discovery necessarily requires a plunge into the unknown. If the answers were already known then there would not be any research to do. Not everyone is cut out to cope with such uncertainty and unfamiliarity, but the ten women discussed here all possessed the curiosity and willingness to attack a blank sheet of paper with gusto and guts. The results they obtained changed the world of science, whether or not their names are familiar in our daily lives.

These women overcame all the many obstacles their gender placed in their way to produce breath-taking results of profound significance, work whose importance still echoes today. We should be grateful to these pioneers and, without sentimentalising their lives, we should appreciate all they did to facilitate the female scientists who have followed in their footsteps. As the L'Oreal tagline puts it 'The world needs science and science needs women'. The lives of the ten women described here provide us with much food for thought and perhaps inspiration for the budding scientists of tomorrow.

Athene Donald, Professor of Experimental Physics, University of Cambridge, and Master of Churchill College

Acknowledgements

We would like to thank the following for their encouragement, forbearance and support: our families; Duncan Proudfoot and his colleagues at Robinson.

Picture credits

....

Introduction

....

'Oxford Housewife Wins Nobel' would not pass muster as a politically correct headline these days. Notwithstanding the inherent sexism in the choice of words, receiving a Nobel Prize does not determine how successful a scientist you are, nor is it something many scientists set out to achieve. Certainly Dorothy Hodgkin, about whom the *Daily Mail* wrote that headline in 1964, was far too busy getting on with the job in hand – probing the structures of complex biological molecules – to be focused on either prizes or headlines. She did not see herself as a feminist or consider too deeply how she was defined. Dorothy was, in her own words, a woman who chose to 'live simply and do serious things'. This was an understatement: she worked extremely hard at a subject about which she was passionate, enjoyed a long and sometimes demanding marriage, had three children, suffered from crippling rheumatoid arthritis and played out a humanitarian role on the world stage.

All in a day's work for Dorothy, so she was a natural choice for inclusion in this book. Others required a bit more consideration. Was it important to choose women who had children in order to suggest that women could have it all? Or was their science the most important thing about them? Delving into their family lives was only one facet of their story in science. And their scientific lives are, after all, the focus here.

In deciding on our ten, we chose women who are no longer alive. That was mainly because distance brings their achievements into focus. Many of these women were not well-known in their lifetime and, in the days before Google existed, finding out about them would have been harder. Nowadays, the news of the 2018 Nobel Prize winner in physics, Donna Strickland – only the third awarded to a woman – was spread worldwide within seconds of its announce-ment. Only one of our choices, Marie Curie, is a household name. We could have chosen to omit her, as she is so well known, but her work on radioactivity was vital, considerably advancing the field of nuclear physics, and she serves as a benchmark for others. Needless to say, the women in this book are more than a match for her.

It's not easy to narrow down to a list of ten, even in the relatively small pool of influential female scientists. Ten seemed a good round number to pick – large enough to provide some breadth but small enough to allow depth too. We have tried to introduce as much vari-ety as possible, to provide a broad picture of the impact these pioneers have made. These women worked in very different areas of science: some lab-based and highly technical, others in medical science or in the environment. There is an international flavour here as well, with American, British, Chinese, Italian and Polish scientists represented.

Not only did these women work in very different areas but they were very different characters too, from the shy Lise Meitner and the retiring but persuasive Rachel Carson to the more outgoing, sociable Virginia Apgar and the strong-willed Rita Levi-Montalcini. It takes all sorts to be a successful scientist. That said, there are some common themes that run through their personalities, science and lives.

All of these women were born within approximately fifty years of each other, with the majority born in the twelve years between 1906 and 1918. As the Victorian industrial age opened up a technological front, they lived through a period of great change, both in the scien-tific world and from a historical perspective. Two world wars, the financial deprivation of the Great Depression, and the Cold War made huge impacts on their lives and working conditions.

Working conditions were often very tough. The name of the game here was exile – from their countries (Lise Meitner), within their countries (Rita Levi-Montalcini) or 'just' from the male-dominated environments of their era, including lecture theatres or the facilities upstairs (Henrietta Leavitt). Where lab space was provided, it was often very cold (Marie Curie) or very hot (Gertrude Elion) or lacking in the most basic health-and-safety measures (Marie Curie and Dorothy Hodgkin). Their work was frequently physically and/or mentally draining.

There were upsides. All of our scientists lived in an age where they didn't have to justify their research in the same way as is required nowadays. No impact statements were necessary. Pure research was just that and they were often much freer to push the boundaries in whichever direction the research led.

Inspiration came frequently, initially in the form of family influence because the parents propagated an intellectual environment and/or because the older generation, particularly the mothers, felt that they had missed out on an education and a career. Family support – both moral and financial – was key, often into adulthood. The home environment and personal experiences also drove some of these women. Rachel Carson's rural idyll and the threat from local industry permeated her later environmental research and writing, while the illness and death of close family and friends propelled Virginia Apgar, Rita Levi-Montalcini and Gertrude Elion into medical research. Later in life, teachers, university lecturers or close colleagues frequently provided the spark that ignited their interest.

Some characteristics are common to all these women: an early, insatiable appetite for learning; persistence – a certain terrier-like mentality; experimental precision; fierce intellectual focus; drive; and intuition. These threads are woven throughout their stories and undoubtedly contributed to their scientific success. It's likely that none of these women would have set much store by analysing the relative merits of these characteristics, nor were they self-publicists. They were too busy doing their science and many, such as Elsie Widdowson, reasoned that no one would be interested in their story. Only one, Rita Levi-Montalcini, wrote an autobiography, wryly titled

In Praise of Imperfection, and that was, in part, achieved because she lived until she was 103 years old!

Many of these women had a very personal approach to science, sometimes at odds with the formal attitudes of their time. Dorothy Hodgkin insisted that everyone in her lab was called by their first names, something that we have adopted for the women in this book. And Dorothy, like Marie and Rita, was not averse to, and indeed encouraged, positive discrimination in her laboratory. Passing on knowledge is a constant theme too, with scientists like Virginia and Gertrude consistently praised for their teaching capabilities and a warm, engaging approach to their students.

It's easy to slip into hagiography when discussing these women but they certainly weren't saints. Chien-Shiung Wu was frequently described as a 'slave driver' by her laboratory staff and her parenting skills were questionable at times. Rita Levi-Montalcini was not known for suffering fools gladly and her argumentative nature frequently got her into trouble. But the facts speak for themselves. These were ordinary women who, often via rather circuitous routes and not without their fair share of mishaps, disasters and family tragedies, did extraordinary things.

A significant number of our scientists had such strong working partnerships that they were married to a particular colleague in their science, and in life in the case of Marie and Pierre Curie. For most, a shared passion did not necessarily extend beyond the laboratory, but some scientific partnerships were so successful that they remained close collaborators for many years, sixty in the case of Elsie Widdowson and Robert McCance. Lise Meitner and Otto Hahn's story was of a more tortuous, unbalanced relationship but, despite this, their joint scientific legacy has stood the test of time. Gertrude Elion and George Hitchings shared an ability to develop a personal relationship with their cancer patients and design radically different drug treatments. Rita Levi-Montalcini and Viktor Hamburger were also equally involved in their science, but partnerships like theirs often encompassed quite different mindsets; Viktor's incremental approach to scientific research complemented Rita's more flamboyant style.

The social context of science is important, and none realised this more acutely than Lise Meitner when she discovered that her research was going to be used to make an atomic bomb. Women like Lise Meitner and Marie Curie were brilliant scientists but they were outsiders, let in reluctantly, often working beyond the walls of the establishment. That put them in a strong position to question and address those issues that others toeing the line could not. Their viewpoint and experience were different. They were open to questioning the processes. Where was the science leading and were the scientists involved guiding that process effectively?

Dorothy Hodgkin was a lifelong promoter of science nationally and internationally. At a time when the Cold War, and the rise of communism, was influencing and hampering scientific research in countries such as China and Russia, she built scientific relationships and kept the lines of communication open. She shared this humanitarian streak with others. Lise Meitner, Marie Curie and Rita Levi-Montalcini worked to help the sick in the two world wars, frequently in distressing and humbling conditions and often using their scientific backgrounds. Rita Levi-Montalcini continued throughout her long life to promote the cause of women, in particular women's education.

Science doesn't work in a vacuum and engaging the public is vital. Many of our ten women were aware of the public interest in science and were keen to reach out to others. Rachel Carson's views sum it up well: 'We live in a scientific age; yet we assume that knowledge of science is the prerogative of only a small number of human beings, isolated and priest-like in their laboratories. This is not true. The materials of science are the material of life itself. Science is part of the reality of living; it is the what, the how, and the why of everything in our experience.'

In an ideal world, a book like this would simply illuminate what interesting scientific lives these women led and any push to correct the gender imbalance and inspire young scientists, particularly women, would fade into the background. For now, we hope that by turning a spotlight on these ten women's experiences of science, and the differences they made to the world, this book will serve as a reminder of what is possible for women in science, with determination, direction and focus.

CHAPTER ONE

Virginia Apgar
(1909–74)

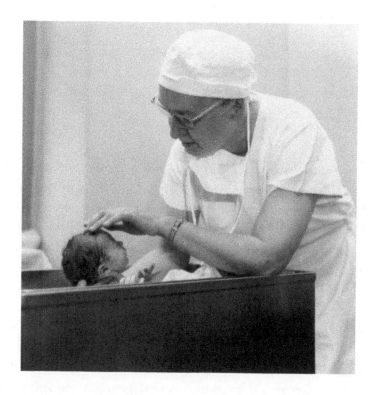

Newborn babies the world over owe their lives to Virginia Apgar and her approach to life, summed up by one of her colleagues as, 'Do what is right and do it now.' Virginia was a trailblazer for modern women in medicine, qualifying in the USA in 1933, when only 5 per cent of doctors were women. She helped to develop the new field of obstetric anaesthetics and her interest in the health of newborn babies resulted in the Apgar test: five simple and quick assessments that can be life-saving and are now used world-wide. The Apgar test laid the foundation for the field of neonatology and Virginia went on to become a world leader in the prevention of birth defects, raising awareness and much-needed funds for research.

Alongside her ground-breaking medical work, Virginia found time for many hobbies, in particular music. She was an energetic, determined and charismatic woman who was undeterred by the chauvinistic nature of medicine and the financial hardships of her early training. Proclaiming that 'women are liberated from the time they leave the womb', Virginia was more than willing to share her joie de vivre and her achievements with others.

Virginia Apgar was born in Westfield, New Jersey on 7 June 1909. The family was delighted to welcome a girl into the family, following the birth of two older brothers, Charles and Lawrence. The Apgar household was a happy, productive and enterprising one. Virginia once said that she came from a family that 'never sat down', a trait that she inherited.

Her father, also named Charles, was a salesman for the New York Life Insurance Company but his real love was science and inventions. He provided an enquiring and creative environment for the young Virginia, sometimes with surprising consequences. Charles was a radio ham and during the First World War he helped to decode messages to German U-boats that were targeting Allied shipping in the Atlantic, and ships were saved.

Creativity was very evident, too, in the form of music, another of Charles's hobbies. The family often held amateur concerts in their living room and both Virginia and her brother Lawrence had music lessons from an early age; Virginia started on the violin at six years old and later the cello, while Lawrence learnt the piano. When the children were old enough, they started performing in front-room concerts and at recitals in local concert halls. Lawrence eventually became a professor of music in Oxford, Ohio.

Virginia's childhood was not entirely free from tragedy. Her eldest brother, Charles Jr, died of tuberculosis just before his fourth birthday, a common occurrence in the 1900s when most of the population were infected and before the advent of antibiotics. Lawrence was two years older than Virginia and suffered from chronic eczema. Like many of her generation, her mother Helen's sole focus was her family, particularly as Lawrence's eczema required much time and effort to keep under control.

Her mother's preoccupation with Lawrence meant that Virginia and her father were able to develop their shared interest in science. It's not clear precisely when or why Virginia decided to become a doctor, but her elder brother's premature death, her father's scientific interests and her mother's caring nature may all have been factors in her decision.

Virginia's drive to achieve academically was helped by her natural intelligence and her affinity for subjects like Mathematics and Greek. At school, she loved debating and was a member of the high-school debating society for four years. With her tall and slim stature, Virginia was a natural athlete and enjoyed tennis and basketball as well. She continued to pursue her interest in music and was a keen member of the school orchestra. Virginia's prodigious energy and involvement in all aspects of school life is reflected in her high-school year book, where her entry ends with the question, 'Frankly how does she do it?'

Studying to degree level, let alone in medicine, was not commonplace for girls when Virginia graduated from high school. Yet she was determined to pursue her interests in science and medicine and, in 1925, at the tender age of sixteen, Virginia enrolled at Mount Holyoke College in South Hadley, Massachusetts. There she specialised in zoology with chemistry and continued to have an active extracurricular life too, playing the violin and cello in the campus orchestra and acting in several plays. She was affectionately known to her peers as 'Jimmy', the girl who did it all. Writing home, unaware of the use of words that would highlight her future work, she reported to her parents, 'I'm very well and happy but I haven't one minute even to breathe.' In 1929, she graduated and set her sights on the next goal – a medical degree.

The timing was not good. In August 1929 the United States economy went into recession, swiftly followed by the stock market crash of October 1929. With the onset of the Great Depression, money was tight for many people and Virginia's family was no exception. Virginia took several odd jobs to support herself, including one in the laboratory of the Zoology Department at Mount Holyoke College. This proved to be an unorthodox occupation by today's standards as her main task was to catch stray cats for the lab, which were humanely killed and preserved for classwork dissections.

With the help of scholarships, the money she earned and some she borrowed, Virginia started at Columbia University's College of Physicians and Surgeons in 1929, at the age of twenty, one of three women in a group of sixty-nine. A fellow medical student, Vera Joseph, who also had to struggle with racial discrimination, remembered her well: 'In her keen, perceptive way, she recognised my need for assurance . . . she would pause for a cheerful greeting, a reassuring hug or a conversation.'

Four years later, in 1933, Virginia graduated fourth in her class and took a surgical post at Columbia Presbyterian Hospital to complete the next stage of her medical training. She impressed her superiors with her skill and intellect but the head of the surgical department, Dr Alan Whipple, dissuaded her from pursuing a career in surgery. He thought that the shortage of posts, particularly for women in the Great Depression, meant that she would struggle to establish herself. And there were the debts she had incurred in her training to consider – almost $4000 – a vast sum, equivalent to over $70,000 today.

With these factors in mind, Whipple was supportive of Virginia's desire to pursue a medical career and suggested anaesthetics as an alternative field. He admired her abilities and spotted the need for training in this new field. In the period 1920–48, no more than 5 per cent of doctors in the USA were women but in anaesthetics approximately 12 per cent were. Other mentors may well have given female doctors advice similar to that of Whipple or perhaps the high percentage was simply because historically anaesthesia had been performed by female nurses.

Anaesthetics was nothing like as advanced as it is today. Although a small number of doctors practised as anaesthetists in the UK in the 1930s, few were specialists in the USA and nurses had been fulfilling this role since the 1880s. In the USA nowadays, 'anaesthetists' are nurses (rather than doctors as they are in the UK) and doctors in this field are called 'anaesthesiologists'. In the 1930s, nurses were often highly competent and technically skilled but many American academic surgeons were concerned about the future of surgery. Surgical procedures were becoming increasingly complex, requiring the development of better anaesthetics, and Whipple suggested that Virginia could make a significant contribution to that field.

In 1934, aged twenty-five, Virginia began her search for a training position, writing to the headquarters of the Associated Anaesthetists of the United States and Canada. Their response revealed a problem: only thirteen training posts were available and only two of these were paid positions. After finishing her surgical post in 1935, Virginia reasoned that it would be better to stay put and learn the fundamentals of anaesthetics from the nurses at Columbia Presbyterian Hospital. In 1937, she spent six months with Dr Ralph Waters in Madison, Wisconsin. He had set up the first academic anaesthetic department in the USA in 1927 and was leading the way in anaesthetic practices.

It was a period of intense learning for Virginia but not an easy one socially. The only woman in her class, she was accepted during the working day, but excluded from dinners and other social events in the evening. Medicine was very much a man's world, exemplified by the lack of housing for female doctors; in her six months in Madison she moved three times. Her housing problem continued when she returned to New York, this time to work with Dr Emery Rovenstine at Bellevue Hospital. Here, she lodged temporarily in the maids' quarters of the clinic.

Undeterred, Virginia finished her anaesthetics training at Bellevue in 1938 and returned to Columbia Presbyterian Hospital. In 1939, aged thirty, she became the second female member of the American Society of Anaesthetists and the fiftieth American doctor to be a board-certified anaesthetist. Soon after, she was elected as its new leader and became the first woman to head the Division of Anaesthesia at the hospital. She set up an organisational structure, establishing residences and incorporated new specialists, without displacing the nurses who had kept the field of anaesthetics alive.

As Virginia's department grew over the next ten years, she expanded her knowledge of anaesthetics, branching out into the new field of obstetric anaesthetics. Her work was now focused on the arrival of new life, but death was also close at hand. In 1950, when she was forty-one, her beloved father died, aged eighty-five; although she mourned his loss, she was grateful that he had witnessed her success as a female doctor. By 1955, that success led to her appointment as the head of the obstetric anaesthesia department at Columbia Presbyterian Hospital.

Her father's influence continued after his death. Virginia's interest in music stemmed from him and, as an adult, even though Virginia was very absorbed in her medical work, she played cello and viola regularly in the Teaneck Symphony of New York and the Amateur Music Players. Her cello and/or viola even accompanied her on her travels, so she could practise and play with local chamber music groups when her busy schedule allowed it.

Not content with just playing her instruments, in 1956 Virginia embarked on another musical odyssey – making her own instruments. She was inspired to do so after meeting a patient, Carleen Hutchings, a fellow music enthusiast who started teaching Virginia all she knew about constructing musical instruments. It wasn't easy finding the time and it wasn't always a quiet occupation either. Virginia's neighbours were kept awake into the early hours as she hammered away in her bedroom, now filled with woodworking tools and her workbench. Virginia's dedication to her new craft resulted in a violin, a mezzo violin, a cello and a viola.

Virginia had a dedicated approach to all her activities, whether it was pursuing her medical career and her focus on the care of newborn babies or in her hobbies. Keeping active was important to Virginia. At school she had played team sports like basketball and later in life she developed a love of golf, angling and gardening. She also enjoyed watching baseball and was a committed fan of the Brooklyn Dodgers.

Virginia's lifelong love of learning, whether it was making musical instruments, learning a new sport or pursing her interests in science and medicine, was often commented on. One of her mentors and good friend, Dr L. Stanley James, described her as 'a student until the day she died. Learning was the focal point of her life. Her curiosity was insatiable and new knowledge held a continuing fascination for her. She was always ready to accept new information and to modify or change her ideas accordingly. She never became rigid. This rare quality enabled her to progress through life without becoming walled in by tradition or custom.'

Her dedication and problem-solving approach to her instrument-making matched her approach to her medical work. Ingenuity was never far from the surface in Virginia's life and the newborn screening test that carries her name, the Apgar test, was equally ingenious in its simplicity and effectiveness.

As Virginia developed her interest and skill in anaesthetics between 1939 and 1949, she was drawn to its applications in obstetric medicine. According to Dr Selma Calmes, a leading anaesthetist,

Virginia entered this field 'in the right time and the right place'. It was not common in the 1940s to be anaesthetised when giving birth, but Virginia's new job involved her in the process of Caesarean delivery. This did require anaesthetic, the application of which was not well understood – in its effect on either the mother or the baby – and resulted in unacceptably high maternal mortality during or just after birth.

Virginia's flexibility and openness to new developments – which extended to her own ability to admit mistakes when they occurred – were instrumental in moving the discipline of anaesthetics forward. New anaesthetics were being developed and Virginia made numerous careful observations of how they affected both the mother and the newborn baby. Between 1949 and 1952, she gathered data on the early moments of a newborn's life, health and prognoses.

Virginia's powers of clinical observation and her ability to make important changes in obstetric anaesthetics, and disseminate that knowledge, did not go unnoticed. One of her colleagues, Dr Stanley James, later said, 'Virginia was not just a doctor. She was also an educator.' As obstetric anaesthetics was such a new field, there was not much published material but Virginia was pragmatic by nature and improvised with what teaching aids were available: she used old bones, or even her own pelvic bone which had an unusual shape, shocking one Australian doctor who had heard about Dr Apgar's pelvis and presumed it belonged to the old and much used skeleton.

The feature of delivery rooms that struck Virginia so forcibly in the late 1940s and early 1950s was the care of babies just after they were born. The focus was more on the mother's wellbeing than the baby's. Hospital deliveries were replacing home births, which meant that more mothers and babies were surviving the birth process, but the first twenty-four hours of a baby's life were still an uncertain time.

There was no routine examination of the newborns' vital signs and, if there was, the methods varied from hospital to hospital and were often unscientific, and even unsafe. Doctors were missing signs that a baby was, for example, starved of oxygen, a factor in half of newborn deaths. Some doctors assumed that babies that were underweight or struggling to breathe should be left to die. 'It was considered better not to be aggressive. You dried them, you shook them and some doctors patted them on the backside and that was it,' said Professor Alan Fleischman, professor of paediatrics at the Albert Einstein College of Medicine in New York.

There was a dire need for a system that checked vital signs, such as heartbeat and breathing rate, from the minute a baby was born. That way, the appropriate special care could be put into place before it was too late.

Virginia's eureka moment occurred one morning while she was having breakfast in the hospital canteen. One of her medical students asked her how to evaluate newborn babies' wellbeing. Virginia replied, 'That's easy. You would do it like this', and jotted down the five vital signs to look for. Initially called the Newborn Screening System, it was the first version of what became the Apgar test.

The medical student may have been surprised by the seemingly instantaneous production of a new scoring system, but Virginia's thoughts were the result of her many years of painstaking observations and clinical knowledge. As a practising anaesthetist, Virginia's daily work involved close contact with newborns. She had seen seemingly healthy babies being whisked away from their mothers to be weighed and measured, only to turn blue and struggle to breathe.

Virginia understood the importance of checking the five vital signs at one minute after birth. Each of these is given a score of zero, one or two. A total score of seven to ten is considered normal, while four to six necessitates some intervention to stimulate, for example breathing, and a score below three leads to emergency treatment. Few babies get a perfect ten one minute after birth, as the circulation – oxygenated blood – often hasn't reached the fingers and toes fully, so these can still be blue.

As the Apgar score became more widely used, it was clear that having a second set of measurements would reveal how the baby was thriving after being born. Comparisons between scores taken at different times would enable the healthcare professionals to monitor improvement or deterioration. Now, the Apgar test is routinely performed at one and five minutes after birth. If necessary, it can be repeated at ten minutes.

After testing the scoring system on more than a thousand newborns, Virginia's test was presented at a meeting in 1952 and, in 1953, Virginia published her findings. The sole-author publication, in *Current Researches in Anaesthetics and Analgesia*, detailed the score's value as a predictor of newborn survival. This is a rare example of a universally accepted and applauded test that was quickly and routinely applied. On the back of the wave of optimism following the advent of antibiotics, the world of medicine was ready for new advances.

Ten years later, in 1963, the newborn screening test was officially designated the Apgar test, when Dr Joseph Butterfield at the Children's Hospital in Denver suggested an acronym using Virginia's surname to help users remember what to look for: **A**ppearance, **P**ulse, **G**rimace, **A**ctivity and **R**espiration. When he wrote to Virginia about it, she replied, 'I chortled aloud when I saw the epigram. It is very clever and certainly original.' Dr Joseph Butterfield published this acronym in the *Journal of the American Medical Association*:

- **A** is for Appearance. A normal pink <u>skin colour</u> all over scores two. If the colour is normal on the body but blue on the feet and/or hands, the score is one. A blue or very pale body is zero.
- **P** is for Pulse. A newborn baby's pulse (<u>heart rate</u>) should be over 100 beats per minute, denoting a score of two. Less than 100 beats per minute is one. If the pulse is absent, the score is zero.

- **G** is for Grimace. This tests for <u>reflexes</u> after stimulating the soles of a baby's feet. If a tickle or gentle slap elicits a jerking movement of the legs or a cough/sneeze, the score is two. A grimace scores one and no reaction zero.
- **A** is for Activity. <u>Muscle tone</u> is indicated by free and regular movement of the arms and legs, scoring two. If the arms or legs are flexed, the score is one. Lack of movement scores zero.
- **R** is for Respiration. If the baby cries and is <u>breathing</u> well after birth, this scores two. If the breathing is slow or laboured, the score is one. Absence of breathing is zero.

Apgar Scale (evaluate at 1 and 5 minutes postpartum)

Sign		2	1	0
A	Activity (muscle tone)	Active	Arms and legs flexed	Absent
P	Pulse	>100 bpm	<100 bpm	Absent
G	Grimace (reflex irritability)	Sneezes, coughs, pulls away	Grimaces	No response
A	Appearance (skin colour)	Normal over entire body	Normal except extremities	Cyanotic or pale all over
R	Respirations	Good, crying	Slow, irregular	Absent

It's easy to see that a high score, with mostly twos, indicates a thriving baby. If the score is mainly ones, some intervention is required, such as extra oxygen. Performing the test again at five minutes after birth then allows doctors to see whether that intervention has made a difference.

Improving interventions, such as the resuscitation of newborn babies if they were not breathing on their own, became a focus for Virginia. Ever aware of the baby's needs, she helped start a revolution in resuscitation strategies. By using less-invasive ventilation and fewer drugs, according to Dr Richard Polin, professor of paediatrics at Columbia Presbyterian Children's Hospital of New York, these improved methods have led to the reduction of chronic lung disease in newborns.

Virginia's life-saving skills were now reaching the wider public and even celebrities. In the winter of 1958, her university magazine, the *Mount Holyoke Alumnae Quarterly*, excitedly reported,

> You never know where '29 [the year Virginia graduated] will turn up next. An Associated Press story last August quoted showman Mike Todd as giving special credit to our Jimmy [a nickname from medical student days] Apgar for her work in saving the life of the premature baby born to his wife, Liz Taylor, screen star. Todd said Dr Virginia Apgar 'worked over the baby for 14 minutes before she hollered. Those were the longest 14 minutes of my life.' . . . She breathed life into the tiny infant.

Keeping people alive was paramount – newborn or adult – and Virginia would entertain friends with resuscitation stories. She always carried a pen knife, an airway tube and plasters in her hand-bag, just in case she had to perform an emergency tracheostomy. This is an opening created at the front of the neck so a tube can be inserted into the windpipe (trachea) to bypass an obstruction and help someone breathe. She once said that she had performed this procedure on sixteen victims of car crashes, saying, 'Nobody, but nobody, is going to stop breathing on me.'

Public health statistical models were changing and, as the Apgar test was rolled out across the USA, it became the gold standard for measuring a newborn's wellbeing. Virginia expanded her research to look at the effect of labour, delivery, anaesthetics and oxygen deprival on the condition of newborn babies. Virginia's colleagues, including Dr Stanley James, helped her with the latest specialist knowledge of cardiology and new methods of measuring levels of oxygen and anaesthetic.

Together the team demonstrated that babies with low levels of blood oxygen and highly acidic blood had low Apgar scores. Lower scores were also associated with certain methods of delivery, types of anaesthetic given to the mother and newborns deprived of oxygen. Virginia noticed that the administration of one anaesthetic, cyclo-propane, to the mother hampered her baby's breathing after birth. This led to the removal of this anaesthetic from the delivery room and the development of epidurals, local anaesthetics that allow women to remain conscious and communicating while giving birth.

The Apgar score was a vital and informative result for one baby; now the cumulative data on thousands of babies revealed some important correlations. Before this simple but efficient scoring system had been invented, doctors, if they had noticed the correlations, did not have sufficient data to prove them. The Collaborative Project, a twelve-institution study involving 17,221 babies established that the Apgar test, particularly the five-minute score, can predict neonatal survival and neurological development. As the Apgar test was more commonly used, babies who needed care started to get it, leading to the development of special-care baby units and newborn-sized heartrate monitors and resuscitation aids.

In 1959, after twenty-six years of dedicated work as a doctor, Virginia decided to take a sabbatical. Although she had a break from the busy life of a medic, she was still focused on helping others, particularly as the Apgar test had stimulated research into the diagnosis and treatment of babies who were born with problems. There was plenty of data emerging on Apgar scores, and their relationship with newborns' health, and Virginia was driven by a desire to pursue this. At the age of fifty, she returned to education and studied for a master's degree in public health at the Johns Hopkins University in Baltimore. This facilitated a change in career direction, into the area of birth defects. Virginia had seen many babies born with these, often reflected in low Apgar scores at birth, and wanted to see how she could improve their immediate and more long-term care.

In January 1938, the National Foundation for Infantile Paralysis had been set up, principally to help children with polio. At the time there was no vaccine and over 50,000 people a year in the USA were paralysed or died from polio virus infection. President Franklin Roosevelt (president in 1933–45) was a polio sufferer, often confined to a wheelchair, and he helped to start the organisation. It later became known as the March of Dimes, a pun based on the newsreel 'The March of Time' and coined by the comedian Eddie Cantor, who was involved in fundraising.

Cantor suggested that, 'The March of Dimes will enable all persons, even the children, to show our President that they are with him in this battle against this disease. Nearly everyone can send in a dime, or several dimes. However, it takes only ten dimes to make a dollar and if a million people send only one dime, the total will be $100,000.' It grabbed the American public's attention and, a month after the launch of the first appeal, 2,680,000 dimes, or $268,000, had been donated to combat polio.

After providing funds for the successful development of a polio vaccine by Dr Jonas Salk, the organisation now focused its efforts on fundraising for research into birth defects, the causes and prevention of prematurity and providing information for parents and the general public. Recent campaigns have included: more effective genetic screening, prevention of spina bifida through folic acid supplementation, and fighting the rising incidence of premature births.

When Virginia gained her MSc in Public Health in 1959, she joined the March of Dimes and worked with the organisation for fifteen years. Very quickly she became more than just a member, as she was made director of the Division of Congenital Malformations. She had decades of experience and her empathetic manner made her a natural choice for the post. Eight years later she became director of Basic Research and by 1973, at the age of sixty-four, she was senior vice president for Medical Affairs.

Virginia was involved in every aspect of the organisation from fundraising to the promotion of health campaigns. Under her stewardship, the March of Dimes grew from a small group to a nationwide organisation, largely due to the amount of money she raised; the organisation doubled its income while she was involved. She felt strongly that her outreach work should focus on reducing the stigma of birth defects and increasing public awareness of the different types of birth defects. Before Virginia's time, parents were encouraged to take babies with birth defects to institutions and relinquish all responsibility for them – an appalling prospect by today's standards but the accepted thinking then was out of sight, out of mind.

The new field of perinatology was growing. This subspeciality of obstetrics is concerned with the care of the foetus and complicated high-risk pregnancies, and is sometimes known as maternal–foetal medicine. Virginia was one of the first people to recognise and inform women about the effect that certain medications or a viral infection could have on their unborn baby.

The potential effect of medications in pregnancy was brought into sharp focus by the thalidomide scandal. In the late 1950s and early 1960s, thalidomide had been given to pregnant women in much of Europe as a treatment for morning sickness and as a sedative. In 1962, its use was linked to 10,000 babies that were born worldwide with missing or incorrectly formed limbs. It was rapidly withdrawn but its effects left a lifelong legacy for those affected babies and their families, and led to enhanced regulation of drugs. Thalidomide was not licensed in the USA, which was played up by the media at the time as a lucky escape, but the message struck home.

The USA was also undergoing a post-war baby boom and new parents were hungry for information about their babies. Virginia travelled extensively, talking directly to parents and doctors alike. One of her major challenges was infections with the rubella virus, which caused thousands of cases of congenital rubella syndrome – premature delivery, miscarriages or still births and also newborn heart problems, blindness and congenital abnormalities, to name but a few. The rubella outbreak in 1964 and 1965 resulted in 20,000 birth defects and 30,000 foetal deaths, propelling Virginia to win funding and government support for a vaccination programme.

Virginia was also instrumental in trying to find ways to prevent premature birth, the March of Dimes' focus since 2003. Her work for the organisation, over the fifteen years that she worked with them, is epitomised by one of its slogans that she coined in the 1960s: 'Be good to your baby before it is born.' Virginia was always thinking about the needs of mothers, and their focus on their unborn child, and in 1972 she co-authored a book with Joan Beck, entitled *Is My Baby All Right?*. This was one of the first books that explained the causes and treatment of a variety of common birth defects and proposed precautions to help improve a woman's chances of having a healthy baby. It was an immediate success.

Virginia was an attractive personality, well-liked by all who met her. She was much in demand as a lecturer and travelled the world both for work and pleasure. Her effervescent personality spilled over into her way of speaking, at breakneck speed. Translators found their job impossible but somehow the message got across and audiences responded to her enthusiasm. Her interest in so many activities made her an interesting person to be around and this spilled over into the lecture room where she enriched her medical talks with stories and anecdotes. One of her favourites became known as the Phone Booth Caper.

In 1957, Virginia and her friend, Carleen Hutchings, had found the perfect piece of maple wood for the back of a viola that Virginia was making. But there was a problem – the wood was already in use as a shelf in a phone booth in a hospital foyer. Not surprisingly, the hospital refused their request to use it but, undeterred, they carried on with their plan. Lifting the wood from its perch, Virginia and Carleen made an exact replica in cheaper wood and replaced it without anyone noticing. Much as Virginia enjoyed recounting the Phone Booth Caper, it remained a private anecdote until the *New York Times* exposed the story some twenty years later.

Her witty personality came across well on television and she was a favourite of TV hosts. Virginia was a people doctor who loved the company of patients and anyone else she met. One March of Dimes volunteer who worked alongside her commented that 'Her warmth and interest give you the feeling that her arms are around you, even though she never touches you.'

In 1973, she also became a lecturer in the Department of Genetics at the Johns Hopkins School of Public Health, fourteen years after she had become a lecturer in Medicine there. Teaching had always been in her blood. From her first days as an obstetric anaesthetist, Virginia was keen to impart her knowledge, frequently in an informal and relaxed style. Teaching would often take place in the hospital corridor or at a patient's bedside, rather than in a more formal lecture hall. Her influence and standing in the field inspired young doctors to specialise in this area and she relished the warm relationship she had with her student nurses and doctors.

Dr Stanley James later commented on her dedication to teaching, 'Whenever she figured out something new about babies and how to best care for them right after they were born, she made sure to teach her new information to as many doctors as she possibly could. Sometimes this meant giving lectures. Other times this meant making short films to distribute to doctors all over the US.' One such video, filmed in 1964, features Virginia teaching the Apgar technique to a student nurse, and illustrates well the calm, patient and encouraging manner that made Virginia such a successful teacher. At the end of it, Virginia says, 'I think this demonstration shows how easy it is to teach these five points of the scoring systems. This young lady is a student nurse and had never heard of this system until this morning. And I think now she has a very good grasp of it.'

Virginia's bedside manner came to the fore towards the end of her life when her mother was one of her final patients. Virginia had never married; noted by her friends for her appalling cookery skills, she joked, 'I never found a man who could cook.' She lived in the same apartment building as her mother and they saw each other regularly. Virginia nursed her mother until she died on 16 March 1969. Only five years after that, Virginia herself became gravely ill. She had been suffering for several years with cirrhosis of the liver and photographs from that time show her looking gaunt and unwell. At the relatively young age of sixty-five, on 7 August 1974, Virginia lost her battle with the disease and was buried beside her parents in the Fairview Cemetery in Westfield, New Jersey.

Her friend and colleague Dr Stanley James gave a heartfelt eulogy at her memorial service at Riverside Church in New York on 15 September 1974, praising her youthful enthusiasm, integrity and insatiable curiosity combined with an honesty and humility that endeared her to all her colleagues and patients. He commented on her 'extraordinary abilities to get the best out of people without antagonising them and to ferret out the essentials and cut into the core of a problem'.

Virginia died relatively young, but the focus on, and enthusiasm for, her Apgar test and her work on birth defects continued unabated. The development of the Apgar test was revolutionary because it was the first clinical method to recognise the newborn's needs as a patient. Since the 1950s, the test has been used routinely worldwide and is still valuable in a technological age. In 2002, based on a study of 150,000 births, the *New England Journal of Medicine* concluded that the Apgar score 'remains as relevant for the prediction of neonatal survival today as it was almost 50 years ago'.

In 1972, Virginia helped to convene the first committee on perinatal health. After four years of deliberation between the March of Dimes and various American medical associations, a landmark study was produced, entitled *Toward Improving the Outcome of*

Pregnancy. Sadly, Virginia did not live to see its inception, but it set out to improve maternal–foetal health and reduce infant mortality, using a model for the regionalisation of perinatal care in the USA. This model was highly successful and contributed to the marked improvement in neonatal survival rates in the decades that followed. The March of Dimes organisation is still going strong and two subsequent reports, produced in 1993 and 2010, have successfully improved and extended this project.

Thanks to the efficiency of the Apgar test and the associated improvement in neonatal care, neonatal mortality (mortality in the first thirty days of life) in the USA has improved by 88 per cent. The same trend has been observed in the UK, from 29.4 neonatal deaths per 1000 live births in the 1930s to 2.8 per 1000 in 2012.

Modern neonatology was revolutionised by the development of the Apgar score. Since it was published over fifty years ago, various adaptations have been made to it, including: taking into account the gestational age of the newborn, the results from ultrasound scans and any administered interventions. The American surgeon and author Atul Gawande has even developed a ten-point surgical Apgar score, published in 2009, that is based on Virginia's Apgar score and measures outcomes after surgery, in patients older than 16 years. Just like Virginia's newborn score, it helps identify patients at higher or lower risk of death or major complications after surgery, providing opportunities for therapeutic interventions.

Virginia's scoring system and focus on the wellbeing of newborns is now wrapped up with an increasing array of ethical issues. The information from the Apgar score, with the advances in technology, means that doctors have to grapple with choosing between what they should do for a sick infant and what they can do. This is particularly difficult and demanding in the case of premature babies.

Although low Apgar scores occur in babies who grow into healthy adults, large-scale long-term epidemiological studies have revealed some associations with lifelong conditions and very low Apgar scores. For example, a study published in 2006 showed that babies with Apgar scores of one to three at five minutes after birth had a higher risk of epilepsy that lasted into adult life. And a paper, published in *Clinical Epidemiology* in 2009, discusses the associations between low five-minute Apgar scores and increased risk of neonatal and infant death and also with neurological disabilities, including cerebral palsy, epilepsy and cognitive impairment.

Virginia had seen similar associations, writing to a colleague in Germany, 'In fact, it turned out, to my surprise, that the development of neurological disorders was closely correlated to the score at one minute.' She was surprised, as she only ever thought of the Apgar score as a device 'to 1) predict infant mortality and 2) to point out to the physician the need for active resuscitation if the total score was four or less'.

Importantly, the 2009 paper stresses that: the Apgar score was never intended for a long-term projection of health and development; the majority of babies with low Apgar scores grow up to be completely healthy adults; and the absolute risks of long-term disability (less than 5 per cent for most neurological conditions) are low to start with.

With the increased awareness of birth defects, their study, known as teratology, advanced rapidly. Virginia's work at the March of Dimes helped the study and diagnosis of such problems, including the increasingly sophisticated use of ultrasound in pregnancy. In 1975, the year after she died, the annual Virginia Apgar Award in Perinatal Paediatrics was first awarded by the American Academy of Paediatrics, to an individual whose career has had a continuing influence on the wellbeing of newborn babies. Those who have received the esteemed Virginia Apgar Award include, in 1992, Dr Joseph Butterfield – the paediatrician who came up with the acronym for the Apgar test.

Virginia was not only one of the first prominent female obstetricians in the USA, she was also the first female doctor to be elected to some key positions of influence and authority. For example, in 1941, at the age of thirty-two, she was made the treasurer for the Board of the American Society of Anaesthetists, serving in that role for four years. By 1946, anaesthetics was becoming an acknowledged medical speciality with required residency training, and in February 1950, her appointment as a professor in the field was proudly announced in her university magazine.

The *Mount Holyoke Alumnae Quarterly* reported, 'Congratulations to Virginia Apgar as professor of surgery at Physicians and Surgeons. We hear she's the first woman so appointed. How about it, Dr Apgar?' This was swiftly followed with an apology in the May 1950 issue, where the news of her surgical appointment was corrected to a full professor of Anaesthetics. Needless to say, it was still a first, both as the first full professorship in this field and achieving this as a woman.

Virginia received many awards in her life and, in 1995, she became a posthumous member of the National Women's Hall of Fame in Seneca Falls, New York. Founded in 1969, this organisation is dedicated to recognising and celebrating the achievements of great American women. Virginia, like Gertrude Elion (see Chapter Four) and Chien-Shiung Wu (see Chapter Ten) is one of the sixty-eight members in the Science section. Virginia's great-nephew was there to receive the award on her behalf. He was only ten years old when she died, twenty-one years previously, but he remembered her clearly and made a speech detailing her full and productive life.

Around the same time, Virginia's contribution to science and medicine was celebrated in another way. She had collected stamps since she was a child and by the time she died had over 50,000. Virginia would no doubt have been amused by the stamp that was issued in her honour in 1994, twenty years after her death. It was a 20-cent stamp used to send postcards and, in the days before email, Virginia's portrait appeared regularly on people's doormats,

Virginia Apgar was only the third female doctor to feature on an

American stamp. When the stamp was unveiled at the American Academy of Paediatrics meeting in Dallas, Texas, on 24 October 1994, a string quartet entertained the delegates with some of Virginia's favourite music. It was an unusual way to introduce a new stamp but the quartet was no ordinary ensemble: the Apgar Quartet were all playing instruments that Virginia had made, or helped to make – a cello, two violins and a viola.

Not only did she learn to make musical instruments but, just a few years before she died, Virginia took up flying lessons. As Virginia was sometimes the only passenger in a small private plane, she reasoned that she ought to be able to bring the plane down herself, if there was a crisis. Judging by a letter to a Californian friend in July 1970, it doesn't seem to have been an entirely smooth process: 'I seem to have the bad habit of wanting to see where the wheels are going to touch down, which means a direct nose dive on to the runway!! I'm working to overcome this deficiency!'

Her daredevil approach to life, combined with a caring and energetic manner, enabled Virginia to revolutionise the medical approach to, and subsequent health of, the newborn child. She helped people understand that babies with birth defects were not just private family tragedies but an important health problem. With the ongoing use of the Apgar test, it has been said that every baby born today is first seen through the eyes of Virginia Apgar.

CHAPTER TWO

Rachel Carson
(1907–64)

Rachel Carson, the biologist, environmental conservationist and writer, was fearless in her quest to understand what was happening to the natural world around her. Her early beginnings as a writer – she published her first compositions at the age of ten – gradually merged with her interest in natural history. Working in marine biology, her combination of scientific scholarship and literary acumen resulted in the publication of four books: a trilogy about the sea, and *Silent Spring*. Published in 1962, 'No single book on our environment has done more to awaken and alarm the world.'

It was a controversial book when it was published but this passionate account of the effects of pesticides such as dichlorodiphenyltrichloroethane (DDT) is credited with opening the eyes of the world to the potential destruction of our environment. When the book was first published in 1962, *Time* magazine found her conclusions 'patently unsound' but, forty years later, their commentary was more laudatory, acknowledging the political and personal battles Rachel had fought: 'Before there was an environmental movement, there was one brave woman and her very brave book'.

Rachel Carson was born on 27 May 1907 in rural surroundings, near Pittsburgh in Pennsylvania, USA. Her mother, Maria, was smitten with her from day one, describing her third child as a 'dear, plump, little blue-eyed baby . . . unusually pretty and very good'. Rachel's siblings, Marian and Robert Jr, were already in school when Rachel was born and Maria Carson had plenty of opportunity to educate Rachel and immerse her in the world of nature around them.

Maria was a studious woman. Trained as a school teacher, she was also a competent pianist and singer with a local choral group. At one of their concerts, she met the retiring company clerk Robert Carson. After they married, Maria was forced to give up teaching, as married women were not allowed to teach in Pennsylvania in the early 1890s. Robert Carson was no farmer, but he was interested in being part of the booming industrial economy of Pittsburgh. He bought sixty-five acres of land, with the intention of dividing this into lots for sale. The Carson home was a small house – two rooms downstairs and two upstairs. It was a basic set-up, with no indoor plumbing, so water was carried from a spring outside. In winter, the family huddled round the fireplace and in summer the children plunged into the local river to cool down.

Rachel's childhood, despite some physical and financial hardships, was an idyllic one. She was free to roam in the fields and woods and explore her growing passion for wildlife. Describing herself as a 'solitary child', she felt at one with nature and from an early age was fascinated by what she saw around her. As an enthusiastic birdwatcher, Maria encouraged Rachel and taught her all she knew about the importance of preserving the environment around them. This extended to rescuing animals whenever possible, such that it was quite common, for example, for the house to be a home for baby robins after their nest had been destroyed.

In 1913, aged six, Rachel started at Springdale grammar school. Her attendance at school was somewhat erratic, partly because Maria Carson was an extremely competent teacher herself and elected for home schooling on a frequent basis, and partly because Maria was keen for her daughter to avoid the prevalent and often devastating childhood illnesses, such as diphtheria, scarlet fever or polio. Rachel was a natural student, frequently achieving A grades, and an avid reader from an early age. Writing stories became a daily focus and Rachel's desire to write, and her natural ability to do so, became increasingly evident. At ten years old, Rachel had her first story published in the children's magazine *St Nicholas*. With the grand title 'A Battle in the Clouds', she joined the ranks of such famous writers as Mark Twain, Louisa May Alcott and F. Scott Fitzgerald who had been published in the same magazine. By the end of 1919, now aged twelve, four of her stories had been published and she had earnt the princely sum of $10. Her destiny, as a writer, was sealed.

Writing was also a form of escape for Rachel. The little two-bedroom house was now crowded with her sister and her new husband and, after military service in the First World War, her elder brother too. Money was tight and the household sought income wherever they could. Maria taught piano at 50 cents a lesson and the younger members of the family worked in the Pittsburgh industrial and power plants, with their vast chimneys spewing out toxic chemicals, an irony that was not subsequently lost on Rachel, the future environmentalist.

Rachel's initial forays into the world of writing covered many subjects but her first published nature story details a day's hike with her beloved dog Pal, describing their delight at finding the 'jewel-like eggs' of the Maryland yellowthroat bird, in a wood 'carpeted with fragrant pine-needles'. Her love of nature and the wildlife around her was beginning to shine from her writing.

In her last two high-school years, at the nearby Parnassus High School, Rachel continued to excel academically and also found time to play basketball and hockey. Her final year thesis was entitled 'Intellectual Dissipation', a fierce diatribe against laziness, particularly of the mental kind. Her strong sense of self-belief, which was to stand her in good stead when she encountered severe criticism later in life, was already much in evidence. Rachel's parents were delighted when she graduated top of her class and was awarded a scholarship to college.

In September 1925, Rachel started the next chapter of her life: four years at Pennsylvania College for Women (PCW) in Pittsburgh. The polluted city environment, with coal dust constantly billowing from the factory chimneys, was quite different to the rural idyll of her childhood home, a mere sixteen miles away. It was more usual to have a higher education than it had been for her mother's generation, but the college president Cora Coolidge still insisted that her students had to attend social etiquette classes and receive grooming lessons.

College was a financial hardship for Rachel's family. Maria Carson was keen for Rachel to be a full-time student and so Rachel, unlike her peers, did not have a part-time job. Instead, Maria took on more piano students and sold some of her family heirlooms, such as silver and china pieces. Her dedication to Rachel's studies even extended to joining Rachel in the college library at the weekends and helping her with her studies, not something that endeared Rachel to her fellow students. Rachel did not curry favour with her peers and was never part of the in-crowd, although she was a keen and highly competent hockey and basketball player. She was not interested in clothes, partying or boyfriends and was often the butt of practical jokes by her classmates. They pretended she had phone calls when she didn't and put washing powder in her bed.

Rachel started out majoring in English, where she attracted the attention of Grace Croff, an assistant professor who taught composition and who encouraged Rachel, successfully, to submit articles for the student newspaper the *Arrow* and the literary supplement the *Engliocode*. Her first story published in the latter publication gives an indication of her burgeoning interest in the sea. She described in lyrical terms the 'towering waves' and the booming breakers. This was despite the fact that Rachel had not yet encountered the ocean she so enjoyed describing.

In her second year at university, Rachel was obliged to take some science courses. Joining her for the semesters of biology, her friend Dorothy Thompson observed that Rachel's self-reliance was easily mistaken for reserve and that actually Rachel was quite friendly and interactive. Rachel was undergoing an epiphany of her own. She was entranced by their biology teacher, Mary Scott Skinker, an elegant and lively woman who shared Rachel's passion for the natural world. For the first time, Rachel realised that she could witness and write about the natural world as well as try to understand the biology that lay behind her observations.

Like her friend Marjorie Stevenson, who was a history major, Rachel was a firm believer in the power of education to make people think, rather than regurgitate facts. Her biology classes were certainly an adventure, revealing aspects of biology she had never considered before. These experiences made Rachel reassess again her direction in life. Writing was viewed as a respectable profession for a young lady but there were few women scientists in the 1920s in the USA. Undeterred, Rachel felt compelled to switch to biology. She might have appeared reserved to some but she was certainly game for a challenge. Her classmates were less certain of her decision. One commented, 'Anyone who can write as well as you can is NUTS to switch to Biology.'

Rachel was determined to push forward with her new direction in life and she spent her last eighteen months at college studying the necessary science courses to graduate as a biology major. It wasn't an easy time. Two of Rachel's mentors were at loggerheads about the importance of science in the education of their young female charges. Mary Scott Skinker wanted students like Rachel to focus on science, whereas the college president Coolidge thought it was distracting from PCW's principal aim of preparing women to be homemakers.

Rachel worked hard at her studies in her final year. Alongside biology, she took courses in physics and organic chemistry and was delighted to be only three of her class of seventy to graduate with a degree magna cum laude (with great honour, a degree of academic distinction in the USA). She was, however, in debt and couldn't be awarded her degree until she settled her $1600 debt to the college. As the Carsons had very little disposable cash, but plenty of land, they transferred two Springdale lots to PCW as security against the debt, to be repaid in due course.

In spring 1929, aged twenty-two, Rachel was awarded a graduate scholarship to Johns Hopkins University in Baltimore, Maryland, USA. On her first attempt, her application had been unsuccessful as she needed a full tuition scholarship. This time the money was forthcoming and she joined the Zoology department, with the *Arrow* student newspaper noting that, 'The honor of this award is seldom conferred upon women.'

Rachel was ready for a change. Although she still loved roaming the woods and fields near her home, Springdale was, like nearby

Pittsburgh, becoming polluted. Her local river was filled with runoff from the factories and was brown and murky. Rachel longed to see the vast and vigorous ocean she had dreamt of and written about.

Arriving by train from Springdale, Rachel found the accommodation in Baltimore less than ideal. There was only one dormitory at Johns Hopkins and this was reserved for men. She eventually found a room off campus and headed out to the mountains of Virginia to stay with Mary Scott Skinker, who was now both a mentor and a good friend. They spent their time riding and walking and, by all accounts, never stopped talking. Mary was a firm feminist who had even broken off her engagement, believing that it would be impossible to continue her career plans if she was married. Rachel was similarly sceptical about marriage, having seen her brother divorced once and her sister twice.

On their return, Rachel made her first visit to the Atlantic Ocean, to the Woods Hole research laboratories, near Boston, Massachusetts, with their dramatic outlook over the ocean, where she was to spend six weeks immersed in research. Here she began to gather some of the facts and ideas about the sea that would later emerge in one of her bestselling publications, *The Sea Around Us*. There was a congenial atmosphere with none of the usual barriers to integration between men and women. Scientists worked hard, with constant lively discussions, and played hard too. Mary Skinker taught Rachel to swim and they joined in the regular tennis parties and picnics.

For the first time, Rachel was able to observe sea creatures in their natural surroundings and meet scientists with similar interests. Dr R. P. Cowles, a marine biologist from Johns Hopkins, would be one of Rachel's professors. Between them, it was decided that Rachel would do her research project for her master's degree on a particular brain nerve, the terminal nerve, in reptiles such as lizards and snakes. Rachel was pleased to meet someone who knew what they were talking about and could help her with her studies and research. Her teaching in some areas of biology to date had been very patchy. Her comparative anatomy was strong but her genetics were weak. Rachel worked hard to compensate for any shortfall and in her organic chemistry class, where she was one of only two women in a class of seventy, she scored highly and commented, 'I was never so proud of an 85 in my life!'

Success in her studies was tempered somewhat when the Great Depression began in 1929. Jobs were scarce for everybody and research positions, those that were available, were targeted at men. In addition, her family had hit upon hard times again. Robert Carson was not a well man and her parents needed her help so, in 1930, they came to live with Rachel in a house she had rented. It was a practical solution for everyone. Her sister Marian and her daughters also joined them. Rachel now had help with housekeeping, never her favourite occupation, and the family could pool its financial resources. Maria Carson lived with Rachel until Maria's death in 1958. It wasn't an arrangement that would have suited everybody but Maria had sacrificed much in her life to support Rachel and they each recognised the role the other had played.

Throughout her time at Johns Hopkins, Rachel did her bit to prop up the family finances by working each summer as an assistant in the zoology lab. The work was mundane, washing up equipment and setting up student practicals, but it served its purpose and helped her family and she made some inroads into the repayment of her PWC debt. In the second year, her scholarship did not cover the tuition fees and she had to work as a lab assistant at the Johns Hopkins Institute for Biological Research. Part-time study was frustrating and Rachel continually felt she was struggling to complete her degree work effectively. Combined with her time pressures, the work on reptiles was not leading to reproducible results and she was finding potential replacement projects difficult to get off the ground. Eventually, she hit upon the topic of catfish embryos and a detailed study of the specialised kidney structure that develops in embryonic catfish before the adult kidney is fully functioning. After hundreds of dissections and detailed drawings, her thesis was complete. It won high praise from her examiners and she was awarded her master's degree on 14 June 1932.

Now, Rachel was keen to embark on a PhD, but the worsening health of her father Robert and her sister Marian meant that Rachel had to drop out of the PhD programme almost as soon as she had started and return to her part-time teaching work. On 6 July 1935, when Rachel was twenty-eight, her family circumstances changed even more abruptly. Her father died and, distraught at her loss, Rachel found herself once again putting the needs of her family first. At this point, an interesting opportunity presented itself. Elmer Higgins, the division chief of the Bureau of Fisheries, was looking for someone to produce radio scripts for educational broadcasts. He was particularly keen to find someone who understood science but who could present it in an accessible and lively manner. The material she would be covering was research on marine life and the fishing industries, and the series was entitled *Romance Under the Waters*. Rachel was taken on part-time, and moved her family to Silver Spring in Maryland. It soon became obvious to her, her editors and Elmer Higgins that she had found her niche.

Her first piece for the *Baltimore Sun* newspaper highlighted the importance of looking after the shad fish adult, which produced the highly desirable shad roe, as well as the welfare of the fishermen. In a telling reflection of the sexist era in which she wrote, the article was written by R. L. Carson. Elmer Higgins recognised that Rachel was a highly competent writer but concealing her gender was viewed as a necessity in the 1930s.

In 1935, at Mary Skinker's suggestion, Rachel took the civil service exams in zoology and Elmer Higgins subsequently appointed her to the Bureau of Fisheries as a full-time junior aquatic biologist. At $2000 a year, the salary was hardly enough to support her family but, significantly, she was one of only two non-secretarial women working at the bureau and it was her first full-time permanent job. The job was also the beginning of a seven-teen-year career in government, building a network of contacts, which would stand her in very good stead when she eventually became a full-time writer.

As she had done as a child, Rachel continually observed and noted the world around. Her notebooks are filled with impressions, records of sounds, sketches and descriptions. Some of this was used for the factual reports she produced for the Bureau and much of the material was written up in the articles that were published in the *Baltimore Sun* and subsequently her books. Early on, one of the themes that would drive her future writing emerged: the destructive effects of industry and pollution on the environment. 'For three centuries, we have been busy upsetting the balance of nature by draining marshlands, cutting timber, ploughing under the grasses that carpeted the prairies. Wildlife is being destroyed. But the home of our wildlife is also our home.'

Her writing was attracting not only the interest of the general public but also that of senior book editors and Rachel felt that her increasing concerns about the environment should be written up as a book. In early spring 1940, she submitted five chapters of her first book, *Under the Sea-Wind*, to the publishers Simon & Schuster. It was accepted with the comment that it was a 'vision splendid'. Armed with a small advance, Rachel pushed on, writing her first drafts by hand and reading phrases out loud to herself. She was a critical writer: working upstairs in her large upstairs bedroom, she made endless revisions, accompanied by her much-loved Persian cats Buzzie and Kito.

Perhaps these cats gave her a different perspective on writing. Rachel was determined to break the mould and write not from the naturalist viewpoint but as though the reader was the creature itself. While using phrases and emotions that related to human experi-ences, she wanted to avoid seeing the marine world through human eyes. Marking her debut as one of the finest nature writers of the twentieth century, the book was a celebration of ecology and the role each creature plays in the life of the sea. Rachel wrote:

> Time measured by the clock or the calendar means nothing if you are a shore bird or a fish, but the succession of light and darkness and the ebb and flow of the tides mean the difference between the time to eat and the time to fast, between the time an enemy can find you easily and the time you are relatively safe. We cannot get the full flavour of marine life – cannot project ourselves vicariously into it – unless we make these adjustments in our thinking.

Rachel felt strongly that human beings needed reminding that we may view ourselves as the superior species but we share this planet with a wealth of interdependent wildlife that needs protecting for our own and their benefit. *Under the Sea-Wind* described the characters and lifecycles of three different creatures – the sanderling, mackerel and eel – living in the dramatic and challenging environment of the ocean. Rachel had set out to be scientifically literate but accessible and lively enough to entice the general reader. She was still anxious and uncertain about the book's approach but her contact, the journalist and historian Hendrik van Loon, reminded her that 'The whole damn business is a gamble . . . what the public will swallow or not . . . who can tell . . . let us hope this time they proved to be fond of fish.' Maria Carson typed up the completed manuscript and it was submitted to Simon & Schuster for publication.

Rachel needn't have feared. The lay public and her scientific colleagues were equally generous in their praise and Rachel felt vindicated in the approach she had taken. Her joy, however, was short-lived. In December 1941, the Japanese bombing of Pearl Harbor had catapulted the USA into the Second World War and the American public were now focused on the war efforts. The inclination, time and money to spend on reading books, however well written, had waned. The book was a commercial failure, selling only 2000 copies. But Rachel never gave up on *Under the Sea-Wind*: it was her personal favourite and since its first publication in 1941 it has been republished several times. The latest, in 2007, features the original drawings by Baltimore artist Howard Frech.

In March 1942, the Fish and Wildlife Service (FWS), which had evolved from the Bureau of Fisheries and Bureau of Biological Survey in 1940, was partially relocated from Maryland to Chicago; Maria Carson moved with Rachel but it was a short-lived move. By the spring of 1943, the Carsons were back in Maryland. Rachel returned to her desk job at FWS, where she became good friends with the artist Shirley Briggs, and wrote magazine articles for publications such as *Reader's Digest* in her spare time. Rachel's writing now covered an even more diverse range of topics including some material that was emerging about a new pesticide, dichlorodiphenyl-trichloroethane, or DDT. This had been used in the Second World War to kill lice and other disease-carrying insects and was now being sold by DuPont without the backing of scientific research to investigate its long-term use. Rachel was uneasy about the effects of DDT on the environment as it seemed to present a sledgehammer approach to controlling pests. And so began a decade-long quest to collect information on DDT.

By 1946, Rachel was directing the FWS publication programme. With a staff of six, she produced a series of twelve booklets entitled *Conservation in Action* that highlighted the destruction of numerous wildlife habitats and the extermination of entire species of animals. Again, she returned to the idea that man and all other species coexist, such that preserving a habitat has direct benefits for man and wildlife. Rachel and her fellow friends and colleagues, such as Shirley Briggs and another FWS artist Kay Howe, took regular trips to collect samples, take photographs, sketch and make records of various species and their environments. They often presented quite a sight on their expeditions to record water fowl across Massachusetts to North Carolina or travelling to the Parker River Refuge north of Boston. Sartorial elegance was not a priority and their motley collection of 'old tennis shoes, wet and besmudged pants' belied the seriousness of their interest and application.

Whenever she could, Rachel returned to the sea, renting a cottage for a month or so in the summer of 1946 near Boothbay Harbor in Maine. Rachel would have loved to live permanently by the sea but she needed the financial security of the FWS work. Her mother was seriously ill at the end of the 1940s, requiring intestinal surgery, and Rachel herself was hospitalised several times, albeit for relatively minor ailments. Her workload was increasing and, needing help, she took on a new artist and collaborator. Bob Hines eventually became a close friend but expressed some doubts initially at working for a woman. These were quickly assuaged when he realised that 'she was a very able executive, with almost a man's administrative qualities'. His rather backhanded compliment was eventually tempered with the realisation that she had a steely core with an intolerance for poor work practices but an inherent sweetness and understanding of good honest people, whatever their level of intellectual ability. He commented that 'she had the sweetest quietest "no" any of us had ever heard. But it was like Gibraltar. You didn't move it.'

Like many women who were trying to pave the way at that time, Rachel was still the victim of sex discrimination. In 1949, aged forty-two, Rachel was promoted but, although she took on some editorial responsibilities, she was not given the grade level or salary to match the man she was replacing. In an ideal world, Rachel would have devoted all her time and efforts to writing, but she didn't yet feel sufficiently confident in her ability to earn a living from book publishing to give up her day job. Mary Skinker had warned her that life as a woman pioneer was tough and Rachel was certainly not alone in experiencing such difficulties. Mary, her college biology tutor and long-standing friend, was now seriously ill and would die of cancer at the age of fifty-seven. Rachel visited her one last time, flying at short notice to be at her bedside in Chicago. When Mary died, Rachel was bereft but turned to her work and her passion for the sea for both solace and inspiration.

Her daily work at the FWS continued but, behind the scenes, Rachel was planning a second book on the sea, encompassing all aspects of the oceans – natural history, chemistry and geology. Around this time, Rachel met Marie Rodell, a publishing agent, who became a key part of Rachel's eventual success and was instrumental in getting *The Sea Around Us* published. Marie was a forthright individual who was renowned for both her knowledge of the publishing world and her loyalty to her clients. In the book, Rachel described how the seas developed, how islands are born, how plants and animals populate the different habitats. She explained how the global winds, rain and tides affect this watery environment. It was to be a complete scientific survey of what is known about the seas, presented with the awe of discovery.

The research process took eight years and included data from a range of sources, including some collected from submarine warfare in the Second World War. Rachel also took full part in the research, with the ongoing support of Marie and other colleagues such as William Beebe, an oceanographer who was instrumental in helping her win a Saxton Fellowship of $2250. This enabled her to take a month's leave to complete her book. Rachel was a softly spoken woman but her quiet determination, combined with an adventurous streak, meant that she was keen to explore the underwater life of the sea. She used her superb research skills and her government knowledge to find the relevant experts in the field. On a trip with the Miami Marine Laboratory biologists, she strapped on a huge diving helmet, attached weights to her feet and plunged into the depths. The experience, although in stormy unsettled water, left a lasting impression on her and she later wrote of the exquisite colours and the 'feeling of the misty green vistas of a strange, non-human world'.

Rachel's sojourn under the surface was followed by another adventure aboard the FWS research vessel, *Albatross III*. This was a first for a woman and was met with a litany of objections by both the fifty men of the crew and various government officials. Eventually, the outing was sanctioned, provided her literary agent Marie Rodell accompanied Rachel as a chaperone. The pair endured a sleepless first night as they adapted to the noisy environment of a working shipping vessel. But the uncomfortable nature of life at sea was soon compensated for by the sheer volume of interesting sea creatures that were trawled up from the depths every night by the vessel's fishing nets.

The Sea Around Us manuscript was typed in its final form by Maria Carson, now eighty-one years old, and the book was eventually published in July 1951. It had been an exhausting process for Rachel, with late nights and weekend work fitted around her full-time job at the FWS, leaving little time for her regular nature walks, but now the accolades and recognition were flooding in. These included a prestigious Guggenheim Fellowship, equivalent to six months of her FWS salary, selection for the Book of the Month Club (BOMC) and the sought-after National Book Award for non-fiction. The book had struck a chord with the public. Its lyrical presentation of millions of years of geological and natural history was a soothing counterpoint to the destruction of the war years. Only six years after the end of the Second World War, there was a yearning to understand the human condition, human beings' place in the natural world, and the dependence of man on earth and the sea. To this day, *The Sea Around Us* is credited as being one of the most successful nature books ever written.

There were downsides to the book's publicity. Once again, Rachel had to deal with the prevailing sexist attitudes. There was disbelief that a woman could exhibit such a depth of scientific knowledge and be able to present it in such an accessible fashion. A reviewer for the *New York Times Book Review* wrote, in the absence of a photograph of the author, that 'it would be pleasant to know what a woman looks like who can write about an exacting science with such beauty and precision'. Her artist friend Shirley Briggs made Rachel laugh by sketching her view of the public's perception of Rachel: a woman of Herculean proportions, standing in flowing water with a spear in one hand and an octopus in the other. Rachel dined out on this joke by reporting how this image of her potential new employer had sent an applicant, who was underqualified for the position the household was seeking to fill, scurrying out of the door.

Some scientists were critical of Rachel's position in FWS's publication division rather than as a laboratory research scientist. Rachel found this criticism strange. As she pointed out, 'science is part of the reality of living; it is the what, the how and the why of everything in our experience'. Others were more complimentary. Henry Bigelow, oceanographer and director of the Harvard Museum of Comparative Zoology, was hugely impressed with the amount of material she had amassed. The general public was clearly impressed too, captivated by Rachel's ability to blend imagination, mystery and wonder with a wealth of scientific facts and her expert knowledge. Four months after publication, 100,000 copies had been sold and *The Sea Around Us* remained on the *New York Times* bestseller list for eighty-six weeks.

As well as coping with sexist and questioning attitudes, Rachel found the attention and the invasion of her privacy overwhelming at times. She was a shy, private individual with a close circle of family, friends and colleagues. Rachel was 'pleased to have people say nice things about the book but all this stuff about me seems odd, to say the least'. This invasion of her privacy came at a difficult time for Rachel. Not only was she coping with the book's publicity but she was recovering from surgery to remove a breast tumour. After surgery, the tumour was classified as benign but it was a salutary warning of the difficult times ahead for Rachel.

Now that Rachel was financially solvent for the first time in her life, she applied for a sabbatical from her job. She was keen to get on with her next book, her third, which had presented itself to her after she heard a Houghton Mifflin editor recount a story. A group of his friends had come across some horseshoe crabs apparently stranded on a beach in New Jersey and had rushed to return them to the sea. Little did they know that they were disrupting the lifecycle of this crab by preventing them from mating. The editor suggested a guide to educate people on the life along their shoreline was long overdue and Rachel was the person to write it.

Rachel returned to her natural habitat and way of life, striding along the coastlines from Maine to Florida, sensing 'the intricate web of life by which one creature is linked to another, and each with its surroundings'. The power of the sea to soothe and absorb Rachel was to be sorely tested over the coming months. While she was learning to cope with the level of public interest in her, Rachel learnt that her unmarried niece Marjorie was pregnant. No great scandal by today's standards but Rachel's newfound fame attracted enough attention for her to feel that the story needed to be kept under wraps and Marjorie protected at all costs. Rachel told only Marie Rodell and put her family first. In September 1952, Marjorie gave birth to a son, Roger Christie. Although Rachel never had family of her own, Roger added to the list of extended family that she cared for throughout her life.

The previous year, when Rachel's sabbatical was over, she had made the decision to finally hand in her resignation in 1951, aged forty-four. The financial success of The Sea Around Us meant that she could retire from government service to write full time. Now Rachel was a free agent, she felt more able to talk about the environmental issues that were of increasing concern to her. This included direct attacks on the administration under the newly elected Republican president, Dwight D. Eisenhower. Several staff changes, including the dismissal of Albert Day, the director of FWS, incensed

Rachel. In a letter to the *Washington Post*, she suggested that 'the way is being cleared for a raid upon our natural resources that is without parallel within the present century ... Hard-won progress is to be wiped out as a politically-minded Administration returns us to the dark ages of unrestrained exploitation and destruction.'

Meanwhile, her new guide to the seashore, *The Edge of the Sea*, was slowly taking shape. It was a tortuous process, as Rachel strived to find the correct tone and approach. A scientifically accurate exploration of the ecology of the Atlantic seashore was required, but Rachel found the creative side of the process demanding and complex. She understood that an intimate understanding of the subject was necessary before reaching a turning point when 'the subject takes command and the true act of creation begins ... The discipline of the writer is to learn to be still and listen to what the subject has to tell him.' As Rachel did this, she realised that the book needed to be classified by the different coastal environments: the northern rocky shore where the tides dictate the flora and fauna, the mid-Atlantic sandy beaches where the waves are more influential, and the southern beaches of coral where the ocean currents govern the life forms.

Rachel put both her mind and body to the test. She frequently returned to the Maine seashore with her artist friend Bob Hines to gather material for the book. These expeditions in the freezing waters were so prolonged at times that Bob would have to carry Rachel back to the shore, as Rachel's muscles seized up in the cold. The Maine coast was always a favourite place and, now she was financially solvent, Rachel bought land on Southport Island and built a small cottage. In July 1953, Rachel, her mother Marie and her latest cat Muffin moved in.

As they settled into their new surroundings, the Carsons were welcomed by their neighbours. Dorothy and Stanley Freeman were keen naturalists themselves and had much enjoyed reading *The Sea Around Us*. Dorothy, in particular, was to become a firm friend of Rachel's; she was warm and engaging, and the two became regular correspondents. Rachel often found writing a lonely existence, and to find a kindred spirit, particularly one who was also looking after an elderly parent, became a great comfort to her.

The Edge of the Sea started out as a field guide to the shoreline but it took on a life of its own as it neared completion, with all the hallmarks of Rachel's inimitable prose. She described memorable encounters with wildlife: descriptions of her exploration of a tide pool, an inaccessible cave and a lone crab on the shore at midnight provide intimate glimpses of the natural world. In the preface, Rachel explained what she had set out to do. She wanted to understand, for example, the life behind an empty shell found on seashore: 'True understanding demands intuitive comprehension of the whole life of the creature that once inhabited this empty shell; how it survived amid surf and storms, what were its enemies, how it found food and reproduced its kind, what were its relations to the particular sea world in which it lived.'

Rachel's joy at spending time at the seashore and developing a deep understanding of it was matched by the general public's reception. Her fears that she would never live up to the reception and success of *The Sea Around Us* were quelled when her editor and the critics pronounced that 'she'd done it again'. Published in 1955, with illustrations by her friend Bob Hines, *The Edge of the Sea* completed Rachel's trilogy on the sea and established her as its foremost 'biographer'.

Inevitably some of the science is now outdated but it remains a masterpiece of ecological writing. Combining the fields of geology, palaeontology, biology and human history, Rachel found a rhythm and clarity that allowed the scientific message to shine through. *The Sea Around Us* became the launch pad for Rachel's most famous and final book, *Silent Spring*.

As her writing took off, Rachel became involved with other projects including a foray into television, writing the script to an *Omnibus* film about clouds, which she watched on her brother's television on 11

March 1956. Her family was still a large part of her life, so much so that she needed a larger house with room to house her and her ailing mother, now in her late eighties, along with her frequently sick niece Marjorie and Marjorie's son Roger. In 1957, Marjorie died, aged thirty-one, and Rachel adopted Roger, who was five years old. Soon afterwards a longstanding family friend Alice Mullen also died. At times like this, stricken with grief and her increasing family commitments, Rachel was overwhelmed and found it impossible to focus on her writing. 'What is needed,' she wrote to Dorothy, 'is a near-twin of me who can do everything I do except write, and let me do that!'

In the late 1950s, Rachel also returned to the subject of DDT, which she and her colleagues at FWS had been concerned about in the 1940s. DDT was the first modern man-made insecticide. Credited with controlling malaria and typhus in the Second World War, it was rapidly and universally accepted as the new way to control both insect-borne diseases and prevent the destruction of crops. But Rachel argued that the science was not in place. Unleashing DDT on the environment had not been fully investigated and it was quite likely that a domino series of downstream effects would take place.

The Agriculture Department was planning to spray twenty million acres of the southern states of America with DDT to control the fire ant, and a similar large-scale spraying was planned on Long Island, New York, to control the gypsy moth. In the latter case, a lawsuit was filed by the local citizens, as the damage to other insects and local birdlife was becoming obvious. There were also reports of DDT resistance and harmful effects on human health. When Rachel's friend, Olga Huckins, wrote an impassioned letter to the *Boston Herald*, describing the horrible deaths of her songbirds, Rachel realised that her attention was now sharply refocused on a 'problem with which I had long been concerned'.

Rachel gradually immersed herself in the world of pesticides and contacted anyone she could who was knowledgeable in this area. Sixteen years with the FWS meant that she had access to all sorts of information and a network of contacts. One of the Long Island plaintiffs, Marjorie Spock, became a friend and sent her numerous research articles, so many in fact that Rachel hired Bette Haney, a college student, to write summaries of the articles.

In May 1957, the judge in the Long Island case refused to stop the spraying of DDT and the appeal process was unsuccessful, too. Rachel believed that there was 'a psychological angle in all this: that people, especially professional men are uncomfortable about coming out against something, especially if they haven't absolute proof the "something" is wrong, but only a good suspicion. So they will go along with a program about which they privately have acute misgivings.' She reasoned that if her book could describe a positive alternative to the use of chemical pesticides such as DDT, her ideas about the damage they were causing the environment, and the human population, would be more readily accepted.

In late November 1958, Maria Carson, aged eighty-nine, had a stroke and died a few days later. It was a devastating loss for both Rachel and Maria's grandson Roger. Rachel desperately missed her mother's gentle but strong presence in her life but resolved to carry on, reasoning that her mother would have wanted her to persevere; in mid-January 1959 she returned to her work. Rachel had now amassed such a volume of evidence about the deleterious effects of DDT, including most significantly the evidence of the harm to human health, that she felt inhibited at presenting it in a public forum for fear of the backlash she would receive. The bonds between the Department of Agriculture and the chemical industry were strong and some of her contacts were loath to be associated with the explosive material they were sharing.

In April 1959, Rachel began to expose her findings. In a letter to the *Washington Post*, she quoted a British ecologist who had described 'an amazing rain of death upon the surface of the earth'. Rachel explained that the rapidly declining robin population was caused by the effect of DDT on the earthworm, the robin's principal food source.

The effect on the food chain was evident elsewhere too. Bald eagles in Florida were now 80 per cent sterile, attributed to the concentration of DDT residues in their dietary fish. And, she, like many doctors and scientists, also believed that humans were affected. One man had died of leukaemia after a three-week hunting trip where he had sprayed the outside of his tent daily with DDT. In November 1959 the cranberry scare added to the headlines. The herbicide aminotriazole, often used to spray cranberries, had been scientifically proven to cause thyroid cancer in laboratory rats. Conscious of the rising tide of public concern, the Food and Drug Administration (FDA) banned the sale of cranberries that had been sprayed with this pesticide.

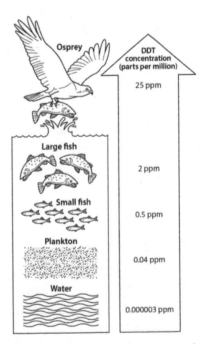

The pesticide DDT becomes concentrated as you go up the food chain

Rachel, and now the general public, were beginning to see the similarities between the effects of radioactivity after the atomic bombing of Hiroshima and Nagasaki in 1945 and those seen with pesticides. This was compounded by nuclear-weapons testing in the 1950s when strontium-90 was released into the atmosphere and purported to filter though the food chain into the US diet, particularly in milk, leading to bone cancers and leukaemia.

Progress on the book was slow. Rachel was keen to present a complete case, commenting to Dorothy, 'This may well outweigh in importance everything else I have done.' Even if she knew something to be true, like the sacking of government biologists because they had questioned the pesticide programmes, she needed documentation that proved this. She was hampered with ill health, too. After an ulcer, pneumonia and a sinus infection she found cysts in her breast, some years after the removal of a supposedly benign tumour. This time, in April 1960, a radical mastectomy was performed and afterwards Rachel enquired whether the growths were cancerous. The answer came back in the negative and no further treatment was recommended.

Rachel's recovery was also slow. She wrote in bed when she could but the book moved at a snail's pace. Her assistant Bette Haney, still not fully in tune with Rachel's meticulous nature, despaired of the work ever being finished. She commented that 'as a child of my culture, I had not yet learned to associate progress with that pace ... and I did not know then the extent of her determination and what a powerful force that kind of determination can be.'

In November 1960, Rachel felt a lump under her ribs near the operation site, and this time radiation and then chemotherapy treatments were recommended. She soon learnt, with the intervention of a friend, Dr George Crile, who was a cancer specialist, that the truth had been withheld from her previously, despite her asking a direct question after her mastectomy. Cancer was a difficult topic of conversation then, even in a doctor's surgery, and doctors were often only willing to discuss a woman's cancer with her husband. Rachel's treatment options were now more limited than they should have been but she held no grudges and resolved to enjoy what time she had.

Roger was her main concern and she wanted to share as much time as possible with him.

At the beginning of 1961, she was weak and often confined to her bed or a wheelchair. Working when she could, she also relished the return of spring, writing, 'All these reminders that the cycles and rhythms of nature are still at work are so satisfying.' Rachel was nearing the end of completing this monumental and revealing book, the culmination of six years' painstaking research into the use of chemical pesticides and their powerful and persistent effects on wildlife and the environment. Marie Rodell had suggested the title *Silent Spring* and Rachel's final task was to write the first chapter: 'A Fable for Tomorrow' describes a harmonious town that gradually fell prey to a 'strange plight'. Death hovered and the animals and birds vanished. Rachel's description of 'a spring without voices' was an amalgamation of a series of true-life environmental disasters, attributed to the pesticide use that she detailed in her book. She warned that, without care, 'this imagined tragedy may easily become a stark reality we all shall know.'

In 1962, the *New Yorker* magazine published a three-part serialisation of *Silent Spring*. It struck a chord with thousands of readers who wrote in, moved and disturbed by not only her vivid account of the poisoning of the environment with toxic pesticides like DDT, but her theory that government and industry were ignoring the outcome. In August 1962, the book was published, heralding the final chapter in Rachel's life. *Silent Spring* became an instant bestseller and the most talked-about book for decades. William Shawn, the editor of the *New Yorker*, whose opinion Rachel greatly valued, described it as a 'brilliant achievement'.

Not all readers were happy with the book and, even after her best efforts to minimise the detractors' influence by sending free copies to prominent government officials and conservation organisations, the backlash began. Rachel was warned early on by a former colleague, 'Facts will not stand in the way of some confirmed pest control workers and those who are receiving substantial subsidies from pesticides manufacturers.'

Rachel was keen to point out that she wasn't totally against pesticide use but just misuse and overuse. She argued that science had portrayed pesticides as a miracle treatment for insect control problems and that any harm they caused was swept under the carpet. Rachel's case was partially saved by another tragedy that couldn't be argued with – the thalidomide scandal. Although it hadn't been licensed for use in the USA, it had been widely prescribed in Europe and Canada to prevent morning sickness, and the startling appearance of malformed babies had struck home. Rachel commented, 'It is all of a piece, thalidomide and pesticides – they represent our willingness to rush ahead and use something new without knowing what the results are going to be.'

Rachel's battles were now in full swing. She was fighting her cancer, which had metastasised, with radiation treatment. And lawyers were gearing up to fight for elements within the chemical industry that were incensed by the tone of her book. The *New Yorker* was asked not to publish further instalments of *Silent Spring* and Houghton Mifflin, Rachel's book publisher, received a letter suggesting that the book was part of a left-wing conspiracy to bring down capitalism, implying that it might be sued if the book were published.

The threat of legal action was deeply worrying and potentially financially ruinous for all concerned, but Rachel and her publisher pushed ahead with support from some in the chemical industry. Just eighteen months into his US presidency, John F. Kennedy publicly acknowledged the importance of *Silent Spring* at a press conference. When asked whether government agencies were examining the potential dangers of pesticides, he said, 'of course, since Miss Carson's book . . . they are examining the matter'. And before long, Rachel was asked to testify before a special presidential committee.

Although the book continued to court controversy, the lawyers backed off – facts were facts, and Rachel was entitled to her opinion and interpretation of the facts she presented. Rachel was somewhat bemused by the furore the book was generating. She had set out to inform the public about the ecology of the natural world and how complex the interactions of living things are. She had cited numerous examples of how introducing the poisonous chemicals contained in pesticides would destroy both the targeted pest and other interrelated organisms in the food chain. Public knowledge of these chemicals and their effects was non-existent and Rachel wanted the public to 'decide whether it wishes to continue on the present road'.

The verbal attacks on Rachel were mainly from those most likely to be affected, for example in the chemical industry, by a potential ban on pesticides or a significant reduction in their use. As finding error in Rachel's reporting of the facts was difficult, these attacks were often highly personal. She was regularly accused of being a communist, dismissive of business and capitalist interests. Others focused on her gender, with a letter to the *New Yorker* saying, 'As for insects, isn't it just like a woman to be scared to death of a few little bugs!' A member of the Federal Pest Control Review Board unkindly commented that 'she was a spinster. What's she so worried about genetics for?'

Rachel's marital status was a recurring theme in the press. When questioned about this, Rachel said that she hadn't had time to get married and that she 'sometimes envied male writers who married because they had wives to take care of them, provide meals, and spare them unnecessary interruptions'. Disregarding any concerns about the lack of a husband, the power of her writing and her understanding of the natural world were second to none.

Silent Spring was chosen as Book of the Month in October 1962. Now, Rachel felt that she was turning the tide. She knew that a Book of the Month book would find its way into homes where reading was not a regular pastime. Furthermore, it would expose how independent research was often not what it seemed, with research frequently funded by the chemical companies and government advisors drawn from an industry background.

Her critics argued that Rachel was not actively involved in research herself until their attention was drawn to her scientific training and the title of her master's thesis: 'The development of the pronephros during the embryonic and early larval life of the catfish (*Inctalurus punctatus*).' Others thought that the good outweighed the bad: pesticides had been phenomenally successful at controlling harmful insects and vermin and keeping certain diseases under control. Rachel's take on pesticide use was viewed as lopsided and, much as her writing style was admired, *Time* magazine viewed the book as 'hysterically overemphatic'.

The response by the chemical industry was instantaneous and defensive. One industry group produced a booklet called 'How to Answer Rachel Carson' and the Monsanto Cooperation published a spoof version of Rachel's first chapter called 'The Desolate Year', which described a world devoid of pesticides. Her scientific friends, who appreciated her analytical mind and knew how much she had strived to make the scientific content of *Silent Spring* accurate, rallied round, allowing Rachel to withstand the barrage of abuse.

For a private person, the publicity surrounding the book was sometimes overwhelming. Rachel switched her telephone number to ex-directory and constantly worried about the press coverage of her. Although this was often scathing, there were lighter moments. The cartoonists had a field day, with numerous interpretations of her concerns for the environment and the cycle of life. For example, one newspaper cartoon shows two men examining a dead dog in the street. One says, 'This is the dog that bit the cat that killed the rat that ate the malt that came from the grain that Jack sprayed.'

Silent Spring was a bestseller both in the USA and abroad. At home, Rachel had agreed to be filmed by CBS. She was in constant pain while they were filming but appeared outwardly calm and engaging. Anyone who watched the broadcast on 3 April 1963, who had not read the book, was now in no doubt that Rachel was alerting the world to the dangers of pesticide abuse, and highlighting that technological advances and the chemical industry were not always at the heart of controlling the world's problems. Six weeks later, the government report from the newly formed special committee review-ing pesticides was published, largely agreeing with Rachel's claims. The CBS follow-up programme concluded, 'Miss Rachel Carson had two immediate aims. One was to alert the public; the second, to build a fire under the government. She accomplished the first aim months ago. Tonight's report by the Presidential panel is prime facie evidence that she has also accomplished the second.'

In the summer of 1963, Rachel knew she was nearing the end of her life but appeared at peace with that thought. With *Silent Spring* published and its messages well and truly heard, Rachel felt she had achieved much of what she had set out to accomplish. Writing to Dorothy from her Maine beach house, she wrote about the Monarch butterflies they had observed: 'For the Monarch, that cycle is meas-ured in a known span of months. For ourselves, the measure is something else, the span of which we cannot know. But the thought is the same: when that intangible cycle has run its course it is a natu-ral and not unhappy thing that a life comes to its end.'

Rachel died on 4 April 1964, aged fifty-seven. Her beloved cats had predeceased her and her major concern had been for Roger's welfare. Money was finally not an issue as there was a significant sum of money in a trust fund. Without consulting either family, perhaps for fear of rejection, she chose two couples – Dorothy Freeman's son and his wife, and her friend and editor Paul Brooks and his wife – as candidates for the future care of Roger. Both families had similarly aged children to the eleven-year-old Roger. In the event, Roger went to live with the Brooks family and the terms of her will, 'a family who would undertake to rear him with affectionate care in the companionship of their own children', were satisfied.

On the day Rachel died, her significance to the world was already being recognised. In Congress, Senator Abraham Ribicoff opened the day's hearing, a committee review of environmental hazards, with the words: 'Today we mourn a great lady. All mankind is in her debt.' These words were proven over the following decades. In 1970, the National Environmental Policy Act promoted 'efforts which will prevent or eliminate damage to the environment and biosphere and stimulate the health and wealth of man'. In the same year, President Nixon established the Environmental Protection Agency (EPA) to coordinate all federal activities that fell under the umbrella of environmental protection. The EPA credits its existence to the seeds of environmental concern that Rachel had sown and, as its website states, the 'EPA today may be said without exaggeration to be the extended shadow of Rachel Carson'. In 1972 DDT was banned for most uses in the USA. A breakdown product of DDT, called DDE, had been shown to thin bird eggshells and reduce their viability. As a direct consequence of the DDT ban, certain bird species such as the bald eagle, the osprey and the peregrine falcon increased in number over the ensuing decade.

Silent Spring has been published in twenty different languages from Arabic to Swedish and, worldwide, the book sales now exceed two million copies. It marked a departure from the tone of her earlier books, with their tender observations of the natural world. It stemmed from, as Paul Brooks described it, Rachel's 'reverence for life' and her increasing concerns that human beings were ignoring and, worse still, destroying the natural order. *Silent Spring* and Rachel Carson's life's work were a wakeup call to the world to preserve the life around us.

It is over fifty years since the book was published but the world is still coping with the effects of industrial pesticides. In June 2013, shoppers in a small town in the USA, Wilsonville, Oregon, discovered more than 50,000 dying bees in a carpark. It was a particularly devastating finding among ongoing concerns about the drastic worldwide drop in these crucial pollinators. It turns out that there was a simple reason. Local workers had sprayed the trees in the vicinity with one of the group of pesticides known as neonicotinoids. Rachel was acutely aware of the mounting chemical threats to bees and throughout *Silent Spring* warns of this. Insecticides are, by their very nature, toxic to bees. In 2012, new scientific findings indicated that some insecticides, namely neonicotinoids and fipronil, showed particularly high risks for bees and, in the European Union, regulations are now in force to ensure that they are used at a low level that minimises their harmful effects to bees.

In Chapter Three of *Silent Spring*, Rachel highlights the dangers of accidental contact with organic phosphate pesticides, often through the reuse of old containers. In July 2013, twenty-three schoolchildren in a village in India were killed or hospitalised after eating a lunch that was contaminated by a pesticide. The pesticide in question, monocrotophos, was mistakenly used to cook the lunch, rather than cooking oil. Monocrotophos has been banned in many countries and is prohibited in India for use on vegetable crops but is still used on cotton.

Rachel was a pivotal figure both as a symbol of the rising environmental movement and as someone whose work led to real change. Her premature death meant that she lived long enough to bear the brunt of the attack on *Silent Spring* but not long enough to witness the transformative practical effects it had. Despite the ongoing use of pesticides, their use has diminished significantly and is now much more heavily controlled worldwide. For example, in the European Union, all pesticides on the market have been subjected to a thorough assessment to ensure a high level of protection of both human and animal health and the environment.

One of the reasons for this decline in pesticides is the use of biological control methods, something that Rachel highlighted in *Silent Spring*. For example, she discussed manipulating insect pheromones to discourage mating, releasing sterilised insects and introducing naturally derived pesticides. Rachel encouraged working with nature rather than against it, and there is now an ever-increasing list of biological control methods at scientists' disposal.

The battle lines are still, to a certain extent, drawn between industry, agriculturalists and the environmental lobby, but *Silent Spring* continues to inform current work on environmental pollution and control. Rachel's work challenged the practices of agricultural scientists and governments, and called for a change in the way mankind viewed the world, a message that is still pertinent today.

CHAPTER THREE

Marie Curie
(1867–1934)

If you were to ask the public to name a female scientist from history, they would probably name Marie Curie. She devoted her life to unravelling the mysteries of radioactivity, discovered two new elements in the process, and paid a heavy price with a long decline into poor health as a result of radiation exposure. Since the introduction of the Nobel Prize in 1901, Marie is the only person, man or woman, to have won it in two different scientific disciplines, physics and chemistry. She spent her career in France, but never forgot her native Poland. Marie survived the tragic death of her husband in a road accident, and a national scandal when she had an affair with a married fellow scientist. She became the first female professor at the Sorbonne and, after her death, the French state had her body moved to the Panthéon to lie alongside the greats of the French nation.

Marie Curie overcame many obstacles in her life, and the first was the accident of where she was born. She was born Maria Salomea Skłodowska on 7 November 1867 in Warsaw, Poland. At the time of her birth, that part of Poland was occupied by the Russians and under their firm control. She was born into an educated family; both of her parents were teachers. Her father, Władyslaw Skłodowski, was a science teacher; it is from him that Marie acquired her love of science and experimentation. Her mother, Bronisława, was the headmistress of a primary school.

In occupied Warsaw, Russian guards patrolled the streets. All things Polish were suppressed, including the language and the teaching of Polish history. All instruction in schools was required to be in Russian, now the official language, and several times a day a guard would drop into lessons to ensure that the Russian-only policy was being kept. Probably due to this oppression, however, the Polish people became defiant and were committed to holding on to their language and traditions. It was commonplace for teachers to instruct in Polish, with a pupil keeping a lookout for any approaching Russian guards. Should one be seen, the teacher would seamlessly switch to Russian until the guard was satisfied, then switch back to Polish once he was out of earshot again.

Marie's parents were as defiant as any of their fellow Poles. Władyslaw positively hated the Russians, and he made no secret of this to his children. In fact, Władyslaw and Bronisława brought up their five children to defy and hate the Russians, too, and to fight for what they believed to be right and just. There is little doubt that this fighting spirit served Marie well throughout her life as she overcame many obstacles that would have defeated others with less determination.

From an early age, Marie adored watching her father going about his work. She was bright and curious, with a particular bent for science. She had frequent chats with her father, which he would skilfully turn into informal lessons. He also encouraged a competitive spirit in his children and would set them weekly mathematics problems. Marie was always thrilled when she was able to solve these, and even more thrilled if she was able to do so quicker than her older siblings.

Władyslaw was not just a scientist, he was also a highly cultured man. He would read poetry to his children and spoke five languages, something he passed on to all of his children. Władyslaw was also a writer and wrote a detailed history of his family in Poland. This was not just an attempt to record his own family's history. It was also a political act to try to preserve the precious Polish heritage and traditions of his family against the backdrop of the Russian occupation.

When Marie started school, she was easily the brightest in the class and she had a phenomenal memory. On one occasion, a Russian guard paid Marie's class a surprise visit. Switching to Russian, the teacher asked star pupil Marie difficult questions, which she answered with flying colours. She felt exhilarated but guilty for showing subservience to a Russian oppressor.

Life got harder for Marie's family over the years. Her mother was diagnosed with tuberculosis and had to quit her job, leaving the family dependent on her father's income. Soon after, her father was dismissed for disobeying his Russian superiors. To support his family, Władyslaw started a school at home; soon there were twenty boys attending classes, some boarding. The Skłodowski house was always crowded and full of activity, an environment in which the ever-curious Marie thrived.

Unfortunately, the home school was also a source of disease. Marie's sister Zosia caught typhus and died, and a few years later her mother Bronisława lost her fight with tuberculosis. Carefree young Marie grew into an introverted teenager who seemed to carry the weight of the world on her shoulders. At the end of the academic year, her teacher recommended that Marie take a break to get over the loss of her mother, but Władyslaw felt that a fresh challenge would be more helpful. He sent her to a tough Russian school where Marie was awarded a gold medal as the best student.

When Marie was fifteen, she was sent to live for a year with her uncle in the country. Away from the influence of her father, she was able to get up late, play outside like a child and indulge in many pursuits that had not been permitted, from fishing and collecting berries to playing games and going to dances. It was possibly the least stressful year of her life.

Back in Warsaw, Marie would have liked to move on to further education but, as was common then, Warsaw University did not admit women. Marie wanted to be a scientist like her father and decided to study on her own, but needed access to laboratories to explore science properly. Luckily, a woman called Jadwiga Dawidowa had started up an unofficial university to teach the women of Poland. Circumventing Russian rules, Dawidowa initially held classes in private homes before moving to bigger buildings, sometimes using school laboratories. Her institution had to keep changing location to avoid discovery. Dawidowa persuaded some of the best academics in Warsaw to give up their free time to teach.

Marie and her sister Bronia attended classes, although both knew that it was a temporary solution. Dawidowa's establishment could not provide recognised qualifications and so Marie and her sister looked elsewhere. The best option would be to study at the Sorbonne in Paris, not only one of Europe's best universities, but one that admitted women. The sisters agreed to take it in turns. Bronia would go to Paris first while Marie stayed behind to work, paying for Bronia's studies and saving for her own future.

Marie became a governess; her first appointment was with what she later described as a 'family of lawyers ... where they don't pay their bills for six months and where they fling money out of the window even though they economise pettily on oil for lamps. They have five servants. They pose as liberals and in reality they are sunk in the darkest stupidity.' Marie hated the experience. 'I shouldn't like my worst enemy to live in such a hell,' she wrote to her cousin Henrietta Michałowska in December 1885.

She left this family and took up a new position in the countryside, starting in January 1886. Marie set off from Warsaw on the fifty-mile journey to the grand house of the Zorawski family, in Szczuki. The Zorawskis had three sons in university, and four children living at home. Marie found the parents to be 'excellent people' and from the start she got on extremely well with her eldest charge, Bronka, who was the same age as her. Her days were busy, but she thrived on the atmosphere in the house. Marie also derived great pleasure from teaching a group of Polish peasant children to read, and continued with her own education, hoping that she was getting closer to enrolling at the Sorbonne.

In January 1888, Marie was in the middle of a personal crisis caused by her growing relationship with her employers' oldest son, Kazimierz. The pair were making plans to marry, but the Zorawskis refused a match between their son and a mere governess, and for fifteen months Marie suffered in silence. In March 1889, she was feeling more buoyant but had still not heard from her sister, so she signed up as governess to the Fuchs family at a resort on the Baltic coast. After a year, Marie finally returned to Warsaw, attending more classes at Dawidowa's university. Bronia then wrote to Marie to tell her that she had finished her studies at the Sorbonne and had married. She invited Marie to Paris to live with her and to start her own studies. Bronia had graduated in the July from the school of medicine, one of three women in a class of several thousand. She knew what was needed to succeed at the Sorbonne and was keen to pass on her knowledge.

So, in November 1891, at nearly twenty-four years old, Marie boarded a train for Paris with a new first name, changing Maria to the French Marie. At the Sorbonne, she was in her element. She excelled at study and revelled in the freedom of her new environment. The only negative was living with Bronia and her husband – their cramped apartment doubled as a surgery and was crowded with patients all day long. After six months, Marie found her own accommodation, closer to the Sorbonne, in the Latin Quarter. She could only afford a tiny room on the top floor of an apartment building, and so badly did she look after herself that, on one occasion, she fainted in the university library. In winter, the water in her washbowl froze overnight. Most nights she slept fully clothed, with the contents of her wardrobe piled on the bed to try to keep warm.

When Marie moved to Paris she knew some French but was far from fluent. She was one of only 210 women at the Sorbonne out of over 9000 students, and among 23 women in the 1825 enrolled in the *faculté des sciences*. Initially, Marie socialised with a small colony of Polish students. By the end of her first year, however, she gave up these friendships to concentrate on her studies.

When Marie arrived, the Sorbonne was undergoing a massive reconstruction. The Third Republic had made it the centrepiece of its educational overhaul of France. Science classrooms and laboratories were still being built, so Marie's classes were held in makeshift quarters nearby while architect Henri Paul Nénot supervised the building of some of the most modern and best-equipped laboratories in the world. The science faculty doubled in size between 1876 and 1900. Money attracted talent. Marie would later write, 'The influence of the professors on the students is due to their own love of science and to their personal qualities much more than to their authority.'

Marie fully intended to return to Warsaw after completing her studies. She envisaged living at home until she met her future husband, but life took an unexpected turn. Graduating top of her class, she was offered a scholarship to stay at the Sorbonne, studying for a *licence ès mathématiques* (the equivalent of a BSc in mathematics). She could not turn this down; science was her life.

In the winter of 1893–4 Marie was also preoccupied with what would become a perennial problem – the search for more laboratory space. She was well along in her preparation for her degree in mathematics and had been hired to study the magnetic properties of various steels. She conducted this research in the laboratory of her professor, Gabriel Lippmann, but it was cramped and poorly equipped. When some Polish friends, Józef Kowalski and his wife, visited her in spring 1894, she moaned about these conditions. Kowalski, a physics professor at the University of Fribourg, Switzerland, knew of someone doing similar research nearby, a Frenchman named Pierre Curie. Pierre had become famous at only twenty-one when he and his brother Jacques discovered that quartz crystals could hold an electrical charge. After that he invented the electrometer, which became the instrument of choice to measure small electrical currents.

When she met Pierre, Marie realised that her life had changed for ever. After the ill-fated affair with Kazimierz, she had devoted her life to study, but Pierre swept her off her feet. She saw in Pierre a kindred spirit and, before long, the two became inseparable. Pierre was different from any man she had known; intelligent and quiet, he loved science as much as she did. Just like Marie's parents, Pierre's family had placed a huge emphasis on education, but he had not followed a conventional path. Pierre had been taught privately at home. At eighteen, he received his *licence ès sciences* from the Sorbonne and was offered a job as an assistant in the Sorbonne's teaching laboratories. Soon he began to publish original work. Pierre had never obtained a PhD, although he had done enough original work to have been awarded several. Without a PhD, he was considered an outsider. In 1893, Pierre left the Sorbonne to teach at a new industrially orientated school, the École Municipale de Physique et de Chimie Industrielles (EPCI).

Pierre was equally bowled over by Marie. He saw in her a startling intelligence. Their friendship quickly deepened, and soon Pierre told Marie that he wanted to marry her. Marie loved Pierre but, after her heartbreak with Kazimierz, she did not want to be hurt a second time. In summer 1894 she decided to leave Paris for Warsaw, feeling that she wasn't prepared to marry. She felt a sense of duty both to her family and to Poland. Pierre was not prepared to give up on this remarkable woman. He wrote, begging her to return to Paris. He even offered to leave France and move to Poland. It is possibly this offer, more than anything else, that made Marie understand Pierre's feelings. She returned to Paris to commit herself to Pierre and a life in France. She did not take up his offer to share an apartment, however, and instead found an apartment on rue de Châteaudun, next to Bronia's office.

Pierre decided finally to write up his doctoral thesis and, in March 1895, he was awarded his PhD and a professorship was created for him at the EPCI. Later that spring, Marie made the final commitment. The pair married on 26 July 1895 at Sceaux town hall in Paris, with the reception in the garden of the nearby Curie family home. Marie's father and sister Helena came from Warsaw and her sister Bronia came with her husband. The Curies used wedding money from a cousin to buy two new bicycles and set off for Brittany on honeymoon, cycling from one fishing village to the next. 'We loved the melancholy coasts of Brittany and the reaches of heather and gorse,' Marie wrote. Returning to Paris after a long honeymoon, the couple took an apartment on the rue de la Glacière, near to where Marie had lived as a student. By this time their combined income from salaries, prizes, commissions and fellowships was about 6000 francs, three times a schoolteacher's salary. This allowed for a comfortable lifestyle but they were not extravagant and did not employ servants.

Marie returned to her work on magnetism and continued to study science and mathematics in her own time. She took two courses, one with Marcel Brillouin, a theoretical physicist with a wide range of interests. Pierre, meanwhile, was given his first class to teach at the EPCI, a course on electricity. According to Marie's later account, the course was 'the most complete and modern in Paris'.

From the start, the couple worked together as often as they could. French mathematician Henri Poincaré once said that their relationship was not just an exchange of ideas but also 'an exchange of energy, a sure remedy for the temporary discouragements faced by every researcher'. In early 1897, Marie discovered that she was pregnant, and suffered frequent dizzy spells and sickness. She was often unable to work and, to add to her low spirits, she learnt that Pierre's mother was terminally ill with breast cancer. Marie feared that her child's birth would coincide with Pierre's mother's death and was worried by the effect this might have on her husband.

Shortly after they returned to Paris from a break in Brittany, Marie went into labour and, on 12 September 1897, gave birth to a daughter, Irène. In Marie's meticulous household expenses book, she noted that they bought a bottle of wine to celebrate. The expenses book also shows a jump in monthly expenditure on employees from 27 francs in September to 135 francs in December. Marie had employed a nurse and a wet nurse for Irène. Then, just as Marie feared, barely two weeks after Irène's birth, Pierre's mother died. Pierre's father moved in with his son, daughter-in-law and new granddaughter.

By now, Marie had finished obtaining the qualification required to be a teacher and, in late 1897, was assembling charts and photographs for her article on the magnetism of tempered steel for the bulletin of the Society for the Encouragement of National Industry. She and Pierre decided that she should begin original research and prepare for a doctorate. The previous two years had been exciting in physics. In 1895, Wilhelm Röntgen discovered X-rays, leaving physicists scrambling to understand this strange new phenomenon. And, in 1896, in what should have been an X-ray experiment, Henri Becquerel uncovered what were first known as Becquerel rays. Marie decided that she would study these new rays for her doctorate, as hardly anything about them was understood.

All that was known was that Becquerel rays were emitted by uranium and that they penetrated paper, making some materials glow in the dark. But, the study of these 'uranium rays' had lost momentum. Only a handful of papers were presented on the subject to the French Academy of Sciences in 1896, compared with nearly one hundred on X-rays. Uranium rays were considered part of the phenomenon giving rise to X-rays; no one realised that they were caused by different processes.

With Pierre's help, Marie set up a laboratory in an old storeroom on the ground floor of the EPCI building. It was cold and dirty, but Marie was happy to devote herself to a research topic of her choosing. She began her laboratory notebook on 16 December 1897. She and Pierre built an ionisation chamber to measure the energy given off by the uranium. These measurements were extremely tricky – Becquerel himself had failed – but Marie believed that with care and diligence she would succeed, and she did.

She began her work with the intention of submitting a PhD dissertation in which she would measure quantities more accurately than before, with no expectations of new discoveries. After testing uranium and measuring the tiny electrical charges in the mysterious Becquerel rays, she scavenged other elements to test. On one day alone in February 1898 she tested thirteen elements, including gold and copper, finding that none gave off uranium rays. Had Marie stuck to testing pure elements, she would have missed the discovery that would make her famous. On 17 February she tried a sample of the black heavy mineral compound pitch-blende. This had been mined for over a century from the mineral-rich Joachimsthal region on the German–Czech border. In 1789, Martin Heinrich Klaproth had extracted a grey metallic element from pitchblende, which he named 'uranium' after the newly discovered planet Uranus.

As uranium was a relatively small component of pitchblende, Marie expected the rays from the pitchblende to be weaker than those from pure uranium. To her amazement, she found the reverse. Initially she thought she had made a mistake, but the result was confirmed after checking. But why was the radiation from pitchblende stronger? Marie tested other substances and, a week later, made another unexpected discovery. The mineral aeschynite, which contains thorium but no uranium, was also more active than uranium. Now she had two puzzles to solve.

Marie suspected that the rays that Becquerel had discovered were not just a phenomenon of uranium, but something more general. The obvious assumption was that there was another, more energetic, element in the pitchblende, also giving off rays. But what could this element be? Pitchblende contained a mix of minerals, too many to duplicate in the laboratory. About this time, the Curies discovered that another uranium-containing mineral, chalcite, also gave off more energetic rays than pure uranium. Chalcite was simpler to synthesise than pitchblende and the Curies reasoned that, if they created chalcite from its known ingredients, the mystery component would be missing and the rays would be less energetic.

Marie mixed up artificial chalcite by combining copper phosphate and uranium, and found that the new mineral showed no greater activity than uranium. The conclusion was clear – chalcite and pitchblende contained an additional unknown component. Marie wrote up her findings as 'Rays Emitted by Uranium and Thorium Compounds'. It was read to the Academy on 12 April 1898. Neither Marie nor Pierre was a member of the Academy and only members could speak but, luckily, Marie's professor, and by now good friend, Gabriel Lippmann, was willing to present the paper. The Academy members were intrigued by Marie's findings but did not pick up on two points in the paper that, in retrospect, were the most important.

PERIODIC TABLE OF THE ELEMENTS

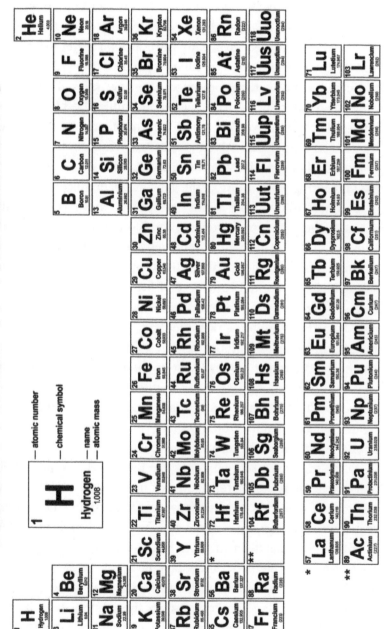

Marie had conjectured that there was a new component in the pitchblende and chalcite that was responsible for the increased energy of the emitted rays. This introduced a novel technique for detecting new elements – that the radioactive properties of a material could indicate their presence. We now know that there are ninety-two naturally occurring elements, but in 1898 far fewer were known. The presence of other, undiscovered elements, was suggested by 'gaps' in the periodic table that Dmitri Mendeleev had introduced to the world of chemistry in 1869, and as each decade passed since his pioneering work, more of the gaps had been filled.

Secondly, in the paper Marie states, 'All uranium compounds are active . . .the more so, in general, the more uranium they contain'. Implicit in this statement is the suggestion that the rays were an atomic property – an idea that turned out to be prophetic. But the Academicians were not convinced of the existence of a new component. The only way to prove this would be to isolate the new component, which could be a new element. This meant dealing with Becquerel. Marie and Pierre were in a difficult position: Becquerel had helped them get money to set up their laboratory and, in many ways, was a friend, but Marie felt slighted by the way that Becquerel always dealt with Pierre and not her. Not only did she feel that he treated her as an inferior, but Becquerel took her ideas and conducted similar experiments in competition to her work.

Pierre tried to reassure his wife that Becquerel was not a rival, but Marie was far more driven than her husband. She didn't want to share her discoveries with anybody else, and certainly not a member of the Academy, a group of men who, she felt, looked down on female scientists. She was determined to beat Becquerel to the discovery of the possible new element. A few days after her paper was read, Marie and Pierre were back in their laboratory, pulverising 100 grams (0.22 pounds) of pitchblende to try to isolate the mystery new element(s). They treated the pitchblende with various chemicals, measuring the activity of the products of chemical reactions. The most active of the breakdown products was then worked on further. Two weeks after starting, the Curies felt that they had isolated enough active product to determine its atomic weight using spectroscopy. This would show conclusively if they had found a new element.

Frustratingly, the substance showed no unknown spectral lines – the bright lines in the light spectrum that act as a fingerprint for identifying elements. This could have meant that Marie was mistaken in her belief in the existence of a new element but, instead, she felt that they had not isolated enough. The Curies asked one of the EPCI's laboratory chiefs, Gustave Bémont, if he could help them in their chemical separation and purification of pitchblende. His input was immediately successful. Beginning by heating a fresh sample of pitchblende in a glass tube, he distilled a product that was strongly active. By early May 1898 they had a substance more active than pitchblende.

By now, the laboratory notebook suggests that Marie and Pierre were dividing their efforts. Very soon, they had a product that was seventeen times more active than the uranium that they used as a benchmark. By 25 June, Marie had a product that was 300 times more active than uranium. Pierre, working in parallel, isolated a product 330 times more active. The Curies were now beginning to think that pitchblende contained not one but two new elements. One seemed associated with the bismuth in the ore and the other with barium. Feeling they had isolated enough of the element associated with bismuth, they tried spectroscopy again, calling upon spectroscopy expert Eugène-Anatole Demarçay, but no new signature was found. Despite this lack of evidence, the Curies were convinced that the bismuth harboured an unknown element. On 13 July 1898, Pierre wrote a significant entry in the notebook. It is the first indication that they had given their hypothetical element a name – 'Po' – an abbreviation for 'polonium', the name the Curies chose in honour of Marie's home country.

Five days later, Henri Becquerel presented a paper on their behalf to the Academy. 'We have not,' they admitted, 'found a way to separate the active substance from bismuth', but had 'obtained a substance which is 400 times as active as uranium'. They went on: 'We thus believe that the substance we have extracted from pitchblende contains a metal never before known, akin to bismuth in its analytic properties. If the existence of this metal is confirmed, we propose to call it polonium after the name of the country of origin of one of us.'

This paper was the first to introduce the term 'radio-active' in its title 'On a New Radio-Active Substance Contained in Pitchblende'. Soon the term radioactivity was everywhere, leaving 'Becquerel rays', 'uranic rays' and 'uranium rays' abandoned.

The aforementioned notebook ends around the time of the paper on polonium, and it seems that they did not get any further for three months. This may have been because they were waiting for a new shipment of pitchblende, but also probably reflected the customary departure of academics from Paris for several months during the *grandes vacances* in the summer.

Once they returned to work, progress was rapid. By the end of November 1898 they had isolated a highly active product, carried off by barium. With Bémont's help, they increased the radioactivity of this product to 900 times that of uranium. This time spectroscopy expert Demarçay found what they had hoped for: distinct spectral lines that could not be attributed to a known element. In late December, Pierre wrote the name for this second new element in the middle of the page of their notebook – radium.

One step remained to prove beyond doubt that they had discovered another element: to isolate it and measure its atomic weight. For several weeks they compared the masses of samples of barium containing radium to the normal element, but they could not detect a difference in mass. They conjectured that this was because radium was present in tiny quantities. At the end of December 1898 they sent off their next paper to the Academy.

Entitled 'On a New, Strongly Radio-Active Substance Contained in Pitchblende', it was written by the Curies and Gustave Bémont. It included the report from Demarçay on his spectral analysis, which not only picked up a new spectral signature but noted that those spectral lines 'intensify at the same time that the radioactivity intensifies . . . a very serious reason for attributing them to the radioactive part of our substance'. Demarçay further added that the spectral lines 'do not appear to me to be attributable to any known element . . . [its presence] confirms the existence, in small quantity, of a new element in the barium chloride of M. and Mme Curie'.

The Curies' laboratory on the rue Lhomond; although short of supplies and equipment, this cold and cavernous space became essential to Marie's success

This announcement marked a turning point in the Curies' work pattern. Rather than continuing to work together on the same project, they set up in parallel. In early 1899, Marie took on the task of isolating radium while Pierre, working in the same laboratory, attempted to understand the nature of radioactivity better. Pierre took on the physics, while Marie handled the chemistry. Marie had a stubborn desire to isolate a sample of radium, but she also knew that to win over the sceptics they had to isolate their new elements.

Marie had to resort to near-industrial methods, requiring a bigger laboratory. The Curies asked the Sorbonne, but all the university could offer was an abandoned building formerly used as a dissection laboratory. The huge space had no heating, so in the winter was horribly cold. Pierre and Marie huddled around a small stove to keep warm, then hurried into the cold parts of the laboratory to conduct their work. There were no extractor hoods to carry away the poisonous gases given off by Marie's chemical treatment, so work had to be done in the courtyard. If the weather did not allow this, they worked inside, opening the windows. By spring 1899 Marie had the materials she needed. As she later recounted: 'I had to work with as much as twenty kilograms [44 pounds] of material at a time ... so the hangar was filled with great vessels full of precipitates and of liquids. It was exhausting work to move the containers about, to transfer the liquids and to stir for hours at a time, with an iron bar, the boiling material in the cast-iron basin.'

Fairly early in the process, it became clear that it would be easier to separate radium from barium than polonium from bismuth. Despite the arduous work and long hours, Marie thrived on the challenge. As she worked on isolating the radium, the Curies had an unexpected source of delight – concentrated radium compounds were spontaneously luminous. Sometimes, after supper, the couple would wander back to their laboratory to admire the eerie glow of their samples. They even sent small amounts of radium to fellow scientists around the world.

Unaware of the dangers, the Curies brought radium salts home in a glass jar, keeping it next to their bed. As the months went on, Marie, Pierre and Becquerel noticed the damage the radioactive materials were causing. Becquerel carried a glass tube of radium salts in his jacket; a few weeks later he found that his skin was burned next to the radium.

Pierre's work was also progressing. He reported the effects of magnetic fields on radium emissions and the Curies published a stream of papers. At the 1900 International Congress of Physics in Paris, they presented their longest paper yet, 'The New Radioactive Substances', in which they summarised their findings, plus work from England and Germany. By this time, it was known that some of the rays could be deflected by a magnet while others could not. Some rays penetrated thick barriers that others could not. And radioactive elements could 'induce' radioactivity in other substances – turning the Curies' laboratory radioactive. But no one knew *how* any of this worked. As the paper said, 'The spontaneity of the radiation is an enigma, a subject of profound astonishment.'

Isolating radium from pitchblende was exhausting and time-consuming. During this long undertaking, Marie presented two progress reports in the journal *Comptes Rendus* in November 1899 and in August 1900. Then, finally, in 1902, Marie announced that she had successfully isolated one decigram (one-tenth of a gram) of radium chloride. Her paper announced that the measured atomic weight of radium was 225, close to the current agreed value of 226, and concluded that 'according to its atomic weight, [radium] should be placed in the Mendeleev [periodic] table after barium in the column of the alkaline earth metals'.

Marie's isolation of radium was not only a huge achievement for sheer doggedness; it was crucial in developing our understanding of radioactivity. Physicist Jean Perrin noted in 1924, 'It is not an exaggeration to say today that [the isolation of radium] is the cornerstone on which the entire edifice of radioactivity rests.'

Marie went on to write up the work for her thesis and submitted it to the Sorbonne. In May 1903 she was awarded a doctorate. The day of celebration also resulted in a chance encounter with Ernest Rutherford, the British physicist, who was in Paris with his wife.

Rutherford had called to see Paul Langevin, who had been a fellow research student in the Cavendish Laboratory in Cambridge in the mid-1890s. Langevin invited the Rutherfords and the Curies to dine with him. After dinner, Rutherford recalled, they went outside into the garden, and Pierre 'brought out a tube coated in part with zinc sulphide and containing a large quantity of radium in solution. The luminosity was brilliant in the darkness and it was a splendid finale to an unforgettable day.'

In August 1903, just two months after receiving her PhD, Marie suffered a miscarriage. Her grief was compounded when she learnt that Bronia's second child had died of meningitis. Marie fell ill with anaemia; it would be months before she could return to work. But, by November, the couple's fortunes began to change dramatically. On 5 November they learnt that the Royal Society of London had awarded them the Humphry Davy Medal, given annually for the most important discovery in chemistry. Marie was too unwell to travel, so Pierre went to London alone. On his return, he found a letter from the Swedish Academy informing the Curies that they, together with Henri Becquerel, had won the 1903 Nobel Prize in Physics.

The Nobel Prize was in its infancy. The first physics prize was awarded in 1901 to Röntgen for the discovery of X-rays. The 1903 prize was to be awarded to Becquerel 'for his discovery of spontaneous radioactivity' and to the Curies 'for their joint researches on the radiation phenomena discovered by Professor Henri Becquerel'. There had been some discussion in the Swedish Academy over whether radioactivity fell within the purview of physics or chemistry. In the end it was considered physics and, to avoid pre-empting a possible future prize in chemistry, the discovery of radium was not mentioned. Marie became the first woman to receive a Nobel Prize and – until her daughter Irène was awarded a Nobel in 1935 – was the only female science laureate.

Pierre thanked the Swedish Academy but went on to say that

neither he nor Marie would attend as they had teaching obligations and important research to conduct, and Marie was not feeling well. Becquerel, on the other hand, went to Sweden, making little mention in his acceptance speech of the Curies' work. Despite this, the newspapers made a great deal of Marie as the first woman to win the prize. The Curies hated the publicity but several good things came out of it. The Sorbonne offered Pierre a professorship and Marie was given a better laboratory. Pierre was also made a member of the French Academy of Sciences.

Meanwhile, more radium was being isolated and the world was falling in love with this strange substance. It was assumed that a substance that gave off such a lovely glow must be good for you, and radium quickly became a cure-all. Some bought radium-enhanced water, while radium salts were sewn into performers' costumes to glow in the dark. In Paris, a Montmartre revue was entitled *Medusa's Radium*, and in San Francisco, USA, a production featured 'fancy unison movements by eighty pretty but invisible girls, tripping noiselessly about in an absolutely darkened theatre and yet glowingly illuminated in spots by reason of the chemical mixture upon their costumes'. Radium was painted onto the faces of watches and clocks, and one company even brought out a radium lipstick. As yet, no one realised the harm the substance could do, but Marie and Pierre were already feeling its ill effects.

Pierre's hands became so damaged from handling radium that he had difficulty dressing himself. His bones ached and he walked like a man twenty or thirty years older. Marie, too, was often weak. It seems strange that neither made any connection between their deteriorating health and the radiation with which they worked. In a paper Pierre wrote during this period, he noted that laboratory animals breathing the emanations of radioactive substances in a confined space died within a matter of hours. The paper concluded, 'We have established the reality of a toxic action from radium emanations introduced into the respiratory system.'

Despite Marie's frequent bouts of weakness and ill health, in December 1904, she gave birth to her second daughter, Ève. The Curies decided to enjoy some of the luxuries that the Nobel Prize money could buy, and started taking more holidays with their children. They had received around 70,700 Swedish krona, the equivalent of around £300,000 ($450,000) in 2017. They bought better clothes and Marie sent a substantial part of the fund to her family in Poland. It seemed as if she had finally found personal and professional happiness.

This was to be shattered on a rainy day in April 1906. Pierre had gone to a meeting and was walking to the Sorbonne. Hampered by the effects of radium, Pierre paused while crossing a busy Parisian street and a large, horse-drawn carriage hit him. Despite the driver's best attempts to avoid him, the carriage wheel ran over Pierre's skull, crushing it instantly. Marie found out the tragic news a few hours later when she got home from work. She was, not surprisingly, devastated.

It was work that slowly helped Marie put aside her grief. The Sorbonne offered her Pierre's professorship, making her the first female professor at the university. Her first lecture, on 5 November 1906, was scheduled to begin at 1.30 p.m., but several hundred people had gathered in front of the Sorbonne's iron gates well before midday to see this historic event. When the doors opened, the crowd rushed in, filling every seat and standing in the aisles. If winning the Nobel Prize made her a celebrity, her determination to carry on with her work led to France taking Marie to their hearts.

Looking for other ways to overcome her grief, she decided to move out of Paris and into the countryside, away from the apartment that carried too many memories. There she arranged private tutors for her daughters. It was already clear that Irène was showing signs of following in her parents' footsteps, exhibiting an early aptitude for science and mathematics. Ève, on the other hand, loved music.

Soon it became known to Marie's friends that she had fallen in love again. The man in question was Paul Langevin, Pierre's former student. However, Langevin was married and his wife discovered love letters from Marie. The wife threatened to kill this other woman. Marie begged Langevin to get a divorce, but he was not prepared to break up his family. He promised his wife that he would not see Marie again, except in a professional capacity. That same year, Marie was nominated as the first female member of the French Academy of Sciences. Yet when the votes were cast in January 1911, she was refused. Her friends were outraged, but Marie shrugged it off.

In November 1911, Marie attended the Solvay conference in Brussels, a meeting for the main players in physics including Albert Einstein, Rutherford, Becquerel, Röntgen and Langevin. Langevin's wife suspected his promise to end the affair was a lie. In a rage, she took Marie's love letters to the newspapers, precipitating a scandal. The day after Marie returned to Paris, the headlines of *Le Journal* carried the front-page headline A Story of Love: Madame Curie and Professor Langevin. The love affair dominated French newspapers for days, almost overshadowing the news that Marie had made history by being awarded a second Nobel Prize, this time in chemistry. The citation was 'for the discovery of the elements radium and polonium by the isolation of radium and the study of the nature and compounds of this remarkable element'. So great was the infamy of the Langevin story that the Nobel Committee wrote to Marie asking her to refuse the prize. Marie wrote back to say that her private life had nothing to do with the quality of her research. She would accept the prize, and in person. In December 1911 she received her Nobel from the King of Sweden, taking her sister Bronia and daughter Irène to the ceremony.

On 29 December, Marie was rushed into hospital and, for the next two years, suffered from a severe kidney ailment. For most of January 1912 she was under the care of the Sisters of the Family of Saint Mary on rue Blomet but, after returning home, her health did not improve and, in March, she went back to hospital for an operation. By this time she weighed only 47 kilograms (103 pounds), 9 kilograms (20 pounds) less than she had three years earlier. She wrote to the Dean of the Sorbonne, asking for time off, and was too ill to return to teaching for another six months. She also suffered increasing bouts of what we now know to have been radiation sickness, but at the time the cause of these bouts was a mystery. Although her affair with Langevin was over, the scandal had taken its toll, and the public who had once adored her did not forgive her readily. She was the scarlet woman who had enticed Langevin into adultery. People threw stones at her windows and newspapers continued to snipe at her.

To stay abreast of developments in her field, Marie travelled under a false name, leaving her daughters with a governess. Gradually, the press lost interest in the scandalous Madame Curie and moved on. Marie eventually found that, once again, she could move about Paris without hostility. In the summer of 1914, Irène passed her baccalaureate and planned to enter the Sorbonne herself. She and her mother were becoming partners as Irène continued to excel at science – the subject she wanted to study at university. Both assumed that they would one day work together in a laboratory, but world events were to intervene. The French government announced the construction of a dedicated centre for Marie's research but the Radium Institute – later renamed the Curie Institute – was opened just as the First World War broke out.

With Paris under threat, the government announced in August 1914 that 'the radium in the possession of Mme Curie, professor of the faculty of sciences of Paris, constitutes a national asset of great value'. On 3 September, Marie took all the radium that France had accumulated in a lead-lined case to Bordeaux and hid it in a vault in the university. As she travelled back she learned that the German army had retreated and the Battle of Marne had begun. French forces, reinforced by soldiers sent to the front line in taxicabs from

Paris and the British Expeditionary Force, overwhelmed the Germans and Marne was won. Paris was safe – for the time being.

Marie realised that the war could provide new purpose and opportunities – in times of crisis, even the moral transgressions of Sorbonne professors seemed trivial. It could allow her to put the pain of the affair behind her. She also saw a way to help her native Poland, which had become the battleground between Russia and Germany. Some sixteen days after the war had begun, the Russian tsar announced that he intended to give Poland its autonomy. In a letter to *Le Temps*, Marie described this as 'the first step toward the solution of the very important question of Polish unification and reconciliation with Russia'.

As was characteristic of Marie, her devotion to the war effort was huge and complete. The notebooks in which she noted every expense show many entries for charitable donations. There are entries for aid sent to Poland, for national French aid, for 'soldiers', 'yarn for soldiers' and shelters for poor people. In addition, according to her daughter Ève's account, Marie invested the money from her second Nobel Prize in French war bonds, which became essentially worthless. She even tried to contribute her medals, but officials at the Bank of France refused to melt them down.

Marie finally hit upon a way to use her expertise to help out, after a conversation with the eminent radiologist Dr Henri Béclère. He told her that X-ray equipment was scarce and 'when it existed was rarely in good condition or in good hands'. Marie became determined to make X-rays available to wounded soldiers at or near the front. Although she was not a radiologist, she knew how to make X-rays; she wrote to Irène, 'My first idea was to set up radiology units in hospitals, employing the equipment that was sitting unused in laboratories or else in the offices of doctors who had been mobilised.'

This hospital work served as a good training ground; she learnt the rudiments of X-ray examination from Béclère and passed this knowledge to volunteers she recruited. These visits to hospitals made her realise that what was really needed was a mobile post that could carry an X-ray machine and all the associated equipment. Marie found benefactors to donate a car – the French Red Cross and the Union des Femmes de France. Finally, she needed to find the necessary equipment. In October 1914, a second car was donated and Marie decided to try to persuade the army to give her radiology cars official backing. After weeks of her request passing from desk to desk, on 1 November, the minister of war granted permission for her radiology cars to go to the front lines. Marie, Irène, the mechanic Louis Ragot and a chauffeur set off in radiology car number two for the Second Army's evacuation hospital at Creil, 20 miles from the front line at Compiègne, north-east of Paris. Marie made a remarkable contribution to the war effort, equipping eighteen radiology cars that helped examine 10,000 wounded soldiers. By 1916, she had obtained a driving licence and so, when necessary, even drove herself.

Marie also trained and educated others, realising that equipment without the training to go with it was useless. She was asked by the army to conduct a course for X-ray technicians but, following months of difficulties with the military approach, she decided to train nurses instead. She opened a school for female radiologists in October 1916 and, between its opening and the end of the war, the school trained about 150 women. They completed a six-week course and were sent to radiology posts around the country.

Marie's recollections of this period were published in her book, *Radiologie et la Guerre*. The greatest praise within was for Irène, who worked closely with Marie throughout the war and became, at the age of only eighteen, a teacher on the course for female X-ray technicians. This partnership continued in the laboratory throughout the rest of Marie's life. By September 1916, Irène was working on her own as a radiologist in Hoogstade, a small part of Belgium that remained unoccupied by Germans. Incredibly, Irène also managed to obtain her certificates from the Sorbonne, passing all subjects with distinction; mathematics in 1915, physics in 1916 and chemistry in 1917.

In the Treaty of Versailles of 1919, Poland became a sovereign nation for the first time in 123 years and Marie wrote, 'A great joy came to me as a consequence of the victory obtained by the sacrifice of so many human lives.'

After the war, Marie wanted to help heal the wounds that had arisen in the scientific community. She was asked to join the commission on intellectual cooperation of the League of Nations and served on it for over twelve years. She also encountered the ambitious American journalist, Marie Meloney, who wrote to Marie asking for an interview. Expecting to find this great woman of French science installed 'in one of the white palaces of the Champs-Élysées', Meloney found herself face-to-face with 'a simple woman, working in an inadequate laboratory and living in a simple apartment on the meagre pay of a French professor'. Meloney decided that Marie needed her help, and Marie sensed an opportunity to acquire some of the radium that the USA had accumulated.

According to Meloney, Marie said, 'America has about fifty grams of radium. Four of these are in Baltimore, six in Denver, seven in New York.' She went on to name the location of every grain. She then added that her own laboratory 'has hardly more than a gram'. Meloney quickly realised that it would help a great deal if she could get enough money together to donate a gram of radium to Marie's laboratory. The journalist set about attracting the funds, in the process portraying Marie as 'impoverished', which was stretching the truth somewhat.

By June 1921, Meloney's fundraising mission was largely accomplished when she managed to raise over $100,000 to buy a gram of radium. Meloney arranged for Marie to visit the USA in the following May, with opportunities to give lectures, accept honorary degrees and to be presented with a gram of radium by President Warren G. Harding at the White House. Marie did not want to travel so early in the year and she pushed for an October visit. Meloney wrote to the rector of the Academy of Paris, encouraging him to put pressure on Marie to agree to a May visit. In the end Marie compromised, leaving Cherbourg on 4 May 1921 with her daughters on RMS *Olympic*. Meloney had by this time committed Marie to a ten-week stay, during which she would attend many luncheons and dinners and award ceremonies, with a little time off to visit Niagara Falls and the Grand Canyon. Large crowds gathered at the pier to welcome Marie in New York and so began a tour that Marie found both exhilarating and exhausting. She was treated like a celebrity with great enthusiasm, much as Albert Einstein had been a few years earlier – at least in most places.

Despite many colleges and universities awarding Marie an honorary degree, the physics department of Harvard voted not to do so. When Meloney asked retired Harvard president Charles Eliot why, he replied that physicists felt that the credit for the discovery of radium did not belong entirely to Marie, and that she had done nothing of great importance since her husband died. Harvard did however welcome her warmly, so Marie herself probably had no idea of these behind-the-scenes machinations.

On 20 May 1921, Marie Curie attended a reception in the Blue Room of the White House. President Harding presented her with the key to a green leather case containing an hourglass with the 'symbol and volume of one gram of radium' (the actual radium was safely stored in a laboratory). Marie's response was brief; she was tired after the hectic tour and had needed to cancel several engagements due to fatigue. At times Irène and Ève found themselves receiving honorary degrees or medals on their mother's behalf. The press was rife with speculation about Marie's fatigue. It was said that 'small talk' was too much for her; that she was unaccustomed to socialising; and that she was not used to leaving her laboratory. But the main source of her illness was undoubtedly her long exposure to radiation. Marie herself said privately during her tour, 'My work with radium . . . especially during the war, has so damaged my health as to make it impossible for me to see many of the laboratories and colleges in which I have a genuine interest.' The Curies made a sweep of the west before boarding the *Olympic* to return to France and Marie's beloved Institut du Radium.

The Institute came out of the desire of the Pasteur Institute and the Sorbonne to build a laboratory for work on radioactivity. After some infighting, a deal had been made to build two separate institutes. One – funded and run by the Sorbonne with Marie as director – was dedicated to the study of the physics and chemistry of radioactive elements; the other focused on the medical applications of radioactivity. The second was funded and run by the Pasteur Institute and directed by Dr Claudius Regaud, a medical researcher from Lyon. The two buildings were built side by side.

From the time that it opened in 1914, Marie's laboratory employed a remarkably high number of women. In 1931, twelve out of thirty-seven researchers were women. This was in stark contrast to nearly every other science research facility in the world. Elsewhere, the few women who were employed were done so as lowly 'computers'; doing calculations for the men who got to do the real research. Marie's laboratory was almost unique in allowing women to do research alongside the men as equal partners in the work. In 1939, Marguerite Perey, working in Marie's laboratory, discovered the element francium and became the first woman to be elected to the Academy of Sciences, fifty-one years after Marie had been rejected.

The risks of working with radiation became more apparent both in and out of the Institute in the years following the First World War. In 1925, a young woman named Margaret Carlough, who painted luminous watch dials in a factory in New Jersey, USA, sued her employer, the US Radium Corporation. She claimed that her work, which involved using her lips to point her brush, had caused irreparable damage to her health. As the lawsuit progressed, it came to light that nine dial painters from the same factory had died and it was concluded that these deaths were due to radiation. By 1928, fifteen dial painters had died from exposure to radium.

In Marie's laboratory, the effects of radiation were also beginning to take a toll. In June 1925, engineers Marcel Demalander and Maurice Demenitroux died within four days of one another from exposure to radioactive materials for medical use. A radiologist had to undergo a 'series of amputations, of fingers, of his hand, of his arm', while another worker lost his eyesight and several others died after terrible suffering. In November 1925, Irène received a letter from the Japanese scientist Nobus Yamada, who worked closely with her preparing polonium sources at the Institute, to say that he had fainted two weeks after returning home and since then had been confined to bed. Two year later, Yamada was dead.

Although she tried to deny it, it was clear to both Marie and those close to her that her own health was deteriorating. She went to extraordinary lengths to hide it. 'These are my troubles,' she wrote to her sister Bronia. 'Don't speak of them to anybody, above all things, as I don't want the thing to be bruited about.' By the early 1920s her eyes had become weak and she had a near continuous humming in her ears. According to her daughter Ève, Marie tried to keep her poor eyesight secret. She placed coloured signs on her instruments and wrote her lecture notes in huge letters. As Ève wrote, 'If a pupil was obliged to submit to Mme Curie an experimental photograph show-ing fine lines, Marie by hypercritical questioning, prodigiously adroit, first obtained from him the information necessary to recon-struct the aspect of the photograph mentally. Then and then alone she would take the glass plate, consider it and appear to observe the lines.'

In all, Marie had three cataract operations and told Ève, 'Nobody needs to know that I have ruined eyes.' But, even if radiation was ruining her health, Marie did not want to retire. As she said to Bronia in a letter in 1927, 'Sometimes my courage fails me and I think I ought to stop working, live in the country and devote myself to gardening. But I am held by a thousand bonds ... Nor do I know whether, even by writing scientific books, I could live without the laboratory.'

Marie returned one more time to the USA, in 1929, to fulfil a promise, as a group of American women had raised enough money to buy another gram of radium for a new institute in Marie's native Poland. This institute eventually opened in 1932. Marie was presented with a cheque for the radium bound for Poland by President Herbert Hoover but, apart from seeing a few friends, she was too ill to repeat the visits of her previous tour.

Marie's health was rapidly deteriorating and, in January 1934, she joined Irène and her husband Frédéric Joliot on a trip to the mountains of the Savoie. Over Easter she made her last visit to her house in Cavalaire with Bronia. Marie went down with bronchitis and had to cut the holiday short. After five weeks convalescing, she returned to Paris where Ève was waiting for her. Marie was increasingly suffering from fevers and chills. From May 1934, Ève witnessed a rapid decline in her mother. Doctors in Paris saw old tubercular lesions on an X-ray, and suggested that Marie be taken to a sanatorium in the Savoie Alps. On the train, Marie fainted but, when she was carried to a bed in the sanatorium, doctors found no evidence of tuberculosis. A Swiss doctor who examined her blood found 'pernicious anaemia in its extreme form'. At dawn on 4 July 1934, in the peaceful sanatorium in the clear Savoie mountain air, Marie Curie died from the steady accumulation of radiation in her body.

Marie has gone down in history as one of the true greats of science, man or woman. As a female scientist she was a pioneer: the first woman to win a Nobel Prize; the first to become a professor at the Sorbonne; the first to direct a major science research institute; and the first to be buried in the Panthéon. She was also a pioneer in combining motherhood with a full-time career in science, paving the way for countless women who came after her.

CHAPTER FOUR

Gertrude Elion
(1918–99)

Gertrude Elion is not a household name but her discoveries have affected all our lives. She was a pioneering chemist in the pharmaceutical industry, who received the 1988 Nobel Prize in Physiology and Medicine with George Hitchings (pictured here) and James Black for their role in making a number of 'designer drugs'. Although she was the fifth woman to receive this Nobel Prize, she was the first who didn't have a PhD or a medical degree.

A native New Yorker and New York University graduate in chemistry, Gertrude was inspired by personal tragedy, including the loss of her grandfather. She persevered for seven years before breaking through gender barriers that prevented women from entering scientific research. Her revolutionary medicines led to a cure for childhood leukaemia, prevented organ-transplant rejection, gave the world its first effective treatment for gout and the first safe antiviral medicine. Gertrude's successes in drug development opened new avenues of scientific research and benefited millions of patients around the world.

Gertrude Elion was born on 23 January 1918 in New York. Trudy, as she was known to family and close friends, was red-haired and lively from day one. At an early age, she developed a voracious appetite for learning and music, attending her first opera at the Metropolitan Opera House with her beloved father when she was only ten years old.

Robert Elion was descended from a long line of rabbis and had come to the USA from Lithuania when he was twelve, one of the two million Jews who fled persecution in Russia and the Austro-Hungarian Empire. Most of the Jews who settled in New York had arrived with no more than $50 to their name and finding work was a priority. Robert Elion saved hard, working nights in a chemist's, to save enough money to go to college. In 1914, he graduated from New York University's College of Dentistry and went on to become a highly successful dentist with several dental practices, and eventually earning enough to invest in the stock market and in property.

Gertrude's mother was also an immigrant. Bertha Cohen came to the USA from Poland when she was fourteen and married Robert when she was only nineteen years old. Their married life started in a large New York City apartment where Robert also had his dental practice. Gertrude's younger brother, Herbert, was born when she was five years old and two years later, the family moved to the Bronx, to an area called the Grand Concourse. There the children enjoyed the delights of the Bronx Zoo, playing games in an empty lot near their apartment and making trips to the local parks.

Robert Elion had a reputation for being wise and sensible, and fellow immigrants regularly came to him for advice. He was often restless, planning elaborate trips and sharing the route maps and train timetables with Gertrude, but the family's financial circumstances did not permit expensive trips aboard or at home. Robert was careful with money, such that Bertha had to justify all her household expenses, and Gertrude developed an early sense of the value of money. 'Getting a little extra money was like filing a new grant application. You had to have an explanation, and you had to essentially go begging for it. You couldn't just go out and buy something.'

Perhaps because of this, Bertha was keen for Gertrude to be self-sufficient and find a career that was well paid. Gertrude's parents encouraged learning too, and not only for the potential financial rewards of an interesting job. Like Robert, Bertha came from a family of scholars, arriving in the USA after her elder sisters had established themselves there. She attended night school to learn English and, with her practical nature, found a job as a seamstress. Gertrude later said, 'She had no higher education but had the most common sense of anyone I knew, and she wanted me to have a career. So she was always very supportive, at a time when many women of her generation would not have been.'

With her parents' encouragement, Gertrude developed an early love of learning. Any subject excited her interest and she was a voracious reader, particularly lapping up any information about two inspirational scientists: Marie Curie, who won two Nobel Prizes for her work on radioactivity (see Chapter Three), and Louis Pasteur, who developed the germ theory of infectious diseases. Gertrude's favourite book was Paul de Kruif's *Microbe Hunters*, an international bestseller that illustrates the achievements, and often tortuous path, of the early microbiologists. Science wasn't necessarily her preferred area at the time – she was an all-rounder – but she thought this book brought the science to life and made a research career sound exciting. As an adult, Gertrude said that she thought every child should read Kruif's book to get a good idea of what it was really like to be a scientist.

Gertrude's innate intelligence and curiosity drove her forwards, propelling her through her early schooling faster than her peers. By the time she reached high school at the age of twelve, she had skipped several years. This had the advantage of keeping her stimulated but meant that she was out of step with her classmates who, with their burgeoning interest in boys and typical teenage pursuits, had very different interests to hers. Gertrude's focus became her work and she derived great comfort and satisfaction from learning.

Studying all her subjects with application, English and history were two of Gertrude's favourites and her teachers expected her to be a writer or a historian. Science certainly wasn't a particular focus and subsequently Gertrude's teachers were surprised that she became a scientist. Gertrude later commented, 'It was an all-girls school and I think many of our teachers were uncertain whether most of us would really go on with our careers. As a matter of fact, many of the girls went on to become teachers and some went into scientific research.'

As she neared the end of her schooling, Gertrude was undecided about her future. Her father wanted her to study dentistry or medicine but she was put off by the dissection classes. It was her close relationship with her grandfather that was instrumental in setting her on her career path. When Gertrude was three, her grandfather had emigrated from Russia to live with the family. He was a watchmaker and a scholar, speaking several languages, including Yiddish. As he grew older, and his eyesight faded, following his profession became impossible, but his relationship with Gertrude went from strength to strength. They enjoyed long walks in the park, exchanging stories, often in Yiddish.

In 1933, just as she was graduating from high school, at the tender age of fifteen, her grandfather became gravely ill. 'He was taken to the hospital and, after a while, I was allowed to visit him. Seeing him there, I remember how shocked I was at his change in appearance. It was the first time I really understood how awful disease could be.' He had stomach cancer and, as Gertrude watched his slow decline and painful death with horror and great sadness, she had a growing realisation of the direction her life would take. As she identified, 'it was a critical time in life . . . because I was just leaving high school. I had to make some sort decision about my future. It was so dramatic that it made an impression at that critical moment. If it had happened earlier, perhaps it wouldn't have.'

His death was the first of four losses that would have a profound effect on Gertrude over the course of her life. 'When I was fifteen, I already knew from my high school courses that I loved science but that year I was so devastated by my grandfather's death from cancer that majoring in Chemistry seemed the logical first step in committing myself to fighting the disease.' Gertrude was driven by a desire to help save others from going through what her grandfather had to endure. 'I felt very strongly that I had a motive, a goal in life that I could try to do something about.'

Studying her chosen subject didn't prove easy, not because she wasn't academically capable, but because money was now very short. The Wall Street stock market crashed in October 1929, when Gertrude was eleven and, like millions of others, the Elions began to struggle. The world economy, and particularly that of the USA, was plunged into the Great Depression and by the winter of 1933, the USA was at its lowest ebb. Robert Elion, who had invested heavily in the stock market, was forced to declare bankruptcy. Robert's dental practice still provided an income but their straitened circumstances meant that they could not finance Gertrude's way through college and beyond.

There was never any doubt in her mind that she had made the right decision, and her family were right behind her. 'Among immigrant Jews, their one way to success was education ... The person you admired most was the person with the most education. And particularly because I was the first born, and I loved school, and I was good in school, it was obvious that I should go on with my education. No one ever dreamt of not going to college.'

Luckily for Gertrude, Hunter College, the women's arm of the City College of New York, offered free tuition in 1933 and Gertrude eagerly enrolled. Women had only been admitted to City College in 1930 and it wasn't until 1951 that the college became fully co-ed. Higher education at the time was largely restricted to Protestant students, so many Jewish students, and others, attended City because there was no other option. Her brother Herbert also took advantage of a City education and followed Gertrude there to study physics and engineering, later running his own bio-engineering and communications firm.

Gertrude loved the stimulating environment of Hunter College. She found one of her chemistry lecturers, Dr Otis, particularly encouraging because he set up a study group for women who were interested in becoming scientists. They met regularly for a journal club, where they discussed novel scientific discoveries recently published in scientific journals – a common enough feature of today's science degree courses, but perhaps an unusual way of teaching at the time, particularly to the women involved. Dr Otis's commitment to his students made a lasting impression on Gertrude and she later tried to recreate this approach when teaching her own students.

In 1937, aged nineteen, Gertrude was awarded a BSc in chemistry, graduating near the top of her class, a helpful boost to her CV at a time when research jobs were not usually open to women. Female science graduates commonly went on to become nurses or teachers but Gertrude had other ideas. Since the death of her grandfather, she was single-minded in her desire to become a research chemist and find a cure for cancer. And Marie Curie continued to inspire Gertrude; if women like Marie could achieve so much, Gertrude clearly felt she could do the same. Gertrude was one of only six or so chemistry graduates in her class of seventy-five women to become a scientist.

First, she had to find a job and the best route into research then, and now, was with a PhD. Gertrude applied to no less than fifteen graduate schools and, despite her stellar grades, had no success. In one notable interview for a research post, she was told she was too pretty and would be a distraction to the men in the lab. Gertrude was dismayed. 'I almost fell apart. That was the first time that I thought being a woman was a real disadvantage. It surprises me to this day that I didn't get angry.'

Discouraged but not downtrodden, Gertrude looked for a job to earn some money to enable her to study further. It wasn't an easy time to be looking for work. In 1937, the Great Depression's effects were still being felt. Jobs were scarce, and in the USA, just as in the UK, women were not encouraged to work outside the home. From 1937 to 1944, Gertrude had many different jobs and occupations, including a six-week stint at secretarial school – an attempt at gaining acceptable skills for a woman – and a job teaching biochemistry to nurses.

Still living at home, and after a year or so at the Denver Chemical Company in New York, Gertrude had saved enough to enrol at New York University in 1939 for a higher degree – a master's in chemistry. There, she was hugely outnumbered by men. 'I was the only female in my graduate chemistry class but no one seemed to mind, and I did not consider it at all strange.'

To help fund her studies, she continued working, this time teaching chemistry and physics to high-school students. After a day's work teaching, she went to university in the evening. There was no let up. At weekends, Gertrude was often found in an unheated lab, wearing a thick winter coat and the Bunsen burner on for heat. Hard work, combined with her ongoing financial difficulties, did not deter Gertrude. There were much larger worries on her mind, particularly as the second of the four losses that weighed so heavily was just about to occur.

While doing her MSc, Gertrude had met the love of her life, Leonard Canter, who was a brilliant young statistician, also studying at City College. He had plans to work for the investment bank Merrill Lynch and, after a period working abroad, he returned to marry Gertrude. However, their plans for the future were abruptly brought to a standstill when Leonard developed fatal subacute bacterial endocarditis – a bacterial infection of the valves and inner lining of the heart. In 1941, there was no life-saving treatment available, as penicillin, whose structure Dorothy Hodgkin investigated (see Chapter Five), was not used for non-military personal until after the Second World War.

Gertrude was devastated by Leonard's death and resolved to dedicate her life to science. Her brother Herbert later commented that the loss of Leonard was 'a heartbreaker and she never fully recovered . . . no one could match up to Leonard'. Gertrude often referred to the compounds she worked on as her 'children'; certainly her lab staff and her students were like family to her and she nurtured her relationships with them and her brother's family.

Looking back, Gertrude said she didn't set out not to get married; perhaps things might have worked out differently if she did. 'This was a time when women couldn't have both a family and a career very easily,' she said. 'I don't think that's true now. I see women who have both. In those days it would have been very much frowned on for a married woman to be working, or to come back to the lab if she had a child.'

In 1941, at the age of twenty-three, Gertrude graduated with an MSc in chemistry – the only woman in her class. About the time the USA joined the Second World War, in late 1941, a job agency rang

Gertrude to see if she was still interested in doing research work. She later commented, 'Interested??? Of course I was still interested! It was all I ever wanted to do. It wasn't until men went to war though, that they finally found they needed me! War changed everything. Whatever reservations there were about employing women in laboratories simply evaporated.'

In 1942, Gertrude got a job as a food chemist at the Quaker Maid Company, a division of the now defunct Great Atlantic and Pacific Tea Company – the American grocery chain known as A&P. Gertrude was pleased to be in a lab environment, but the work was routine and didn't fulfil her intellectual curiosity. 'I tested the acidity of pickles, the mould in the frozen strawberries; I checked the colour of egg yolk going into the mayonnaise. It wasn't exactly what I had in mind but it was a step in the right direction.'

After eighteen months with A&P, she had a six-month job with the healthcare company Johnson & Johnson. Then, at the age of twenty-six, she had a lucky break, one that set her up for nearly forty years. Her father was given a sample of the painkiller empirin to use in his dental practice. It was made by the pharmaceutical company Burroughs Wellcome (now GlaxoSmithKline) and he wondered whether it would be worth phoning to see if they had any work for a lab assistant. Gertrude was a little dubious but phoned anyway and spoke to a receptionist, who suggested she drop in one Saturday. After all those years, seven years since she first graduated, struggling to even get an interview, it was an easy passage into the world of Burroughs Wellcome.

The pharmaceutical industry often gets a bad name for rating financial gain before more humanitarian ideals. Burroughs Wellcome was different. From the time it was founded in England in 1880 (by Henry Wellcome and Silas Burroughs), the company's philosophy centred on the development of drugs to cure serious diseases. At the time, drug production was in its infancy but, by 1900, the discoveries of Louis Pasteur and Robert Koch, and their fellow scientists, had led to the identification of twenty-one disease-causing microorganisms. Burroughs Wellcome was the first drug company to employ research scientists to develop new therapies and, by 1912, eight branches had opened worldwide, including the one in New York in 1906.

Soon after that fateful phone call in 1944, Gertrude put on her best suit and went to Burroughs Wellcome for her interview with George Hitchings. She was relieved to see there was one other woman working in their lab, out of a staff of seventy-five, although their early relationship was not all plain sailing. Her fellow chemist, Elvira Falco, was initially unsupportive, as Gertrude recounts, 'She told him [George Hitchings] not to hire me because I was too well dressed.' Understandably affronted by this apparent display of jealousy, Gertrude retorted, 'Well, wouldn't you wear your best suit for an interview?' Despite this awkward start, the two women did become good friends.

Thankfully, George Hitchings ignored Elvira's comments and took Gertrude on as an assistant biochemist. Gertrude was still bothered by her lack of a PhD but George saw through that apparent disadvantage. He was already well known for his alternative working style and his decision-making was well respected. Impressed by her obvious intelligence, knowledge and drive, he agreed to her request for a salary of $50 a week, a good rate at the time, commenting that he 'thought she was worth it'.

George and Gertrude did not always agree on everything but their forty-year partnership was close. Throughout their career, George was one step ahead of Gertrude; every time he got a promotion, she stepped into his shoes. When he retired in 1967, Gertrude became the head of the Department of Experimental Therapy, the first woman in the USA to lead a major research group, in what was now one of the world's leading pharmaceutical companies.

Research scientists at Burroughs Wellcome had the freedom to pursue ideas, an environment that suited Gertrude down to the ground. She was eager to learn more and became something of a polymath. Over her long career in science, unlike today's scientists who often specialise exclusively in one area, her work covered a variety of scientific disciplines, including organic chemistry, biochemistry, pharmacology, immunology and virology.

Conditions in her first lab were not perfect. The baby-food plant downstairs dehydrated food all year round, and there was no air conditioning to alleviate the intense heat emanating from the floor, so Gertrude had to wear thick rubber-soled nurses' shoes. But she never felt the need to find another working environment: 'Science is the kind of discipline where you keep learning all the time. I always wanted a job where you didn't stop learning and there was always something new.' By the end of her working career, she was the inventor on forty-five patents for novel therapies, author on more than two hundred papers, and the recipient of twenty-three honorary degrees, not to mention the small matter of a Nobel Prize.

George Hitchings was interested in making the process of drug discovery a more rational and less hit-and-miss affair. His research methods were highly unusual for the time. Most pharmaceutical research was focused on a specific disease, doggedly testing different compounds to see whether they had a therapeutic effect – the so-called trial-and-error approach. George approached the field from the opposite direction and developed a whole new process for research into new drugs, now called rational drug design. This is the method most commonly used today, focusing on understanding the properties of a chemical or compound first, then working out how best to develop this as a therapy.

Today's rational drug design involves pinpointing a particular protein as the target. When the precise 3D structure of the protein is known, the drug can be designed to fit with the shape the structure reveals. In the days before scientists such as Dorothy Hodgkin (see Chapter Five) had revealed the structure of proteins, George

Hitchings didn't have this information at his disposal, but he was fascinated by the idea of anti-metabolites. Investigation of the first type of antibiotics, the sulfonamides, had showed that they blocked an essential metabolite used by the bacteria.

In 1944, when Gertrude joined George Hitching's lab, the work was focused on the metabolism of nucleic acids, the molecules that carry genetic information. Working on these molecules at the time was not an obvious path to new therapies; it was only in 1944 that a scientific paper by Oswald Avery had cautiously suggested that the nucleic acid called deoxyribonucleic acid (DNA) was the key component of genes. And it was another nine years before James Watson, Francis Crick and Rosalind Franklin solved the structure of DNA, revealing how the genetic information might be copied during cell replication.

Notwithstanding their incomplete knowledge, Gertrude and George knew that all cells need nucleic acids to reproduce and that cancer cells, with their uncontrolled growth, require large amounts of nucleic acids – significantly more than other cells – to sustain their replication. Cancer cell growth and replication is rapid too, requiring fast production and repair of DNA, making them vulnerable to compounds that would disrupt this process of replication.

George Hitchings also knew that the large molecules of DNA (and ribonucleic acid, RNA) were made up of simple chemical units – the DNA building blocks called bases. Two of these became Gertrude's speciality: the purines – adenine and guanine. These are two of the four chemical bases found in DNA (which are now known to pair with their complementary pyrimidine bases, thymine and cytosine respectively, forming the helical structure of the DNA molecule). Hitchings proposed that they could make false versions of the purine bases that would inhibit the replication of rapidly dividing cells, such as cancer cells and bacteria.

It was pioneering work. Gertrude commented later that 'few chemists were interested in the synthesis of purines in those days

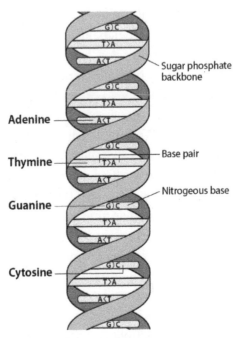

The structure of DNA showing the double helix, with a pyrimidine (thymine or cytosine) base joined to its complementary purine (adenine or guanine) base to form a base pair

and I relied on methods in the old German literature'. George and Gertrude set out to synthesise bases that were similar enough to the natural occurring molecule to integrate into the DNA but dissimilar enough that they would jam the mechanism of replication. These became known as 'rubber doughnuts', antagonists that looked and behaved like the real thing but weren't.

Gertrude used the purines not only as potential drugs but also as research tools that might be able to revealing metabolic pathways that at the time were mere suggestions. As Gertrude commented, 'Let the drug lead you to the answer nature is trying to hide from you.'

The two were instrumental in pioneering rational drug design but this still meant long hours at the bench. 'At the beginning,' Gertrude recalled, 'it was my job to find out how to make compounds.

So, I'd go to the library to see if I could figure out how to do it . . . I would just go ahead and make the compounds, and then the question was, well what do we do with these compounds. How do we find out if they really do anything?' What follows in this chapter is not an exhaustive description of all the drugs Gertrude and George developed but rather a representative set.

The initial challenge was to find a biological method to determine the potential activities of the new compounds. Gertrude and George hit upon a microbial screening system, looking at the growth of *Lactobacillus casei*, a harmless bacterium used for making Cheddar cheese, which can grow on defined mixtures of purines. In 1948, they found that one of the many derivatives of the purines they studied, diaminopurine, strongly inhibited the growth of laboratory cultures of *L. casei*. When tested on mouse tumours, or tumour cells in tissue culture, diaminopurine demonstrated a similar inhibitory effect.

In 1948, Gertrude's first potential cancer treatment was tested on patients at the Sloan-Kettering Institute in New York. After a promising start, the emotional and physical rollercoaster ride of cancer chemotherapy began. Diaminopurine was not well tolerated by many of the recipients and its toxic nature often resulted in intense vomiting. Gertrude watched from the sidelines as the patients went into remission only to relapse two years later.

A key feature of Gertrude and George's research was the personal relationship they developed with many of their patients and how intensely and personally they felt their patients' discomforts. They were involved at every step of the way and were much closer to the action than many of today's drug developers, staying involved as the compounds advanced through the pipeline, from lab to bedside. One of Gertrude's colleagues, Dr Krenitsky, observed that this wasn't uncommon then but 'Trudy was a master at it. She'd argue with the medical people. She'd argue with the FDA [Food and Drug Administration].'

As diaminopurine was first being used in the clinic, Gertrude watched as the third loss that profoundly affected her life occurred. 'JB' was a twenty-three-year-old woman who appeared to be improving with this new drug. After seemingly recovering from cancer, she married and had a child. But the cancer was still lurking and she lost her battle. Chemotherapy regimens and cancer monitoring were very different then and she was given no further drugs after the initial treatment. Today, other drugs would have been administered and/or the dosing regimens adjusted – she may have survived to see her child grow up. Gertrude was deeply affected by the loss of this young woman and resolved to push on with her research, trying to find a drug that would be effective but less toxic to normal cells than diaminopurine.

Gertrude was realistic about the nature of scientific discovery, saying, 'Research is very hard work. There's no other way, but how you handle setbacks can make a difference. In science, you just have to take several approaches to setbacks ... You must never feel that you have failed. You can always come back to something later, when you have more knowledge or better equipment and try again. I've done this and it worked!'

Gertrude's first major clinical success, at the age of thirty-two, was also a drug for cancer, the disease that had propelled her into her research career and which continues to challenge researchers. In the 1940s and 1950s, half of the children who developed acute lymphoblastic leukaemia, a cancer of the white blood cells, died within a few months.

By the early 1950s, Gertrude and George had made and tested over a hundred purines using the *L. casei* bacteria screening. Guanine's structure contains two chemical rings, and substitution of an alternative atom into one of the numbered positions on these rings can alter the properties of the molecule. They discovered that the substitution of oxygen by sulfur at the 6-position of guanine produced an inhibitor of purine utilisation. 6-mercaptopurine

(6-MP) was tested at the Sloan-Kettering Institute, with whom Gertrude and George had established a collaboration, and were found to be active against a wide spectrum of rodent tumours and leukaemias.

After the initial tests for efficacy and toxicity in mice, 6-MP underwent its first successful clinical trial in children. Forty per cent of these children were completely cured. For the first time, a drug had been found that could treat this devastating disease. Many of the patients relapsed at various intervals, but the powerful effects observed led the FDA to approve the drug 6-MP, whose trade name is Purinethol®, in 1953. This was little more than two years after its synthesis and before the FDA had received any comprehensive scientific data on it, a much less thorough process than would have been sanctioned nowadays.

Drug development is now a much longer and less personal process, often taking between ten and fifteen years from beginning to end. After selecting a target gene or protein involved in a disease, researchers search through up to ten thousand candidate compounds to eventually arrive at one approved therapy. After pre-clinical testing in computer models, *in vitro* cell culture and *in vivo* animal trials, there are three phases of clinical trials. Phase I checks for safety issues, such as toxicity, in 20–100 healthy volunteers, Phase 2 is a clinical trial for efficacy in 100–500 patients with the disease, and Phase 3 is a large-scale clinical trial in thousands of patients to gain further information about safety, efficacy and benefit versus risk relationships.

6-mercaptopurine is still in use – one of a group of chemotherapy drugs known as anti-metabolites that stop cells making and repairing DNA. It is not a cure but slows down the course of the disease, allowing other therapies time to work. Using 6-mercaptopurine with other drugs in a combination therapy regime means that the five-year survival rate for acute lymphoblastic leukaemia now exceeds 90 per cent in children under fourteen years old. 6-mercaptopurine is on the World Health Organization's Model List of Essential Medicines – the most important medications needed in a basic health system – and is also now used as an immunosuppressive therapy for Crohn's disease and ulcerative colitis.

Gertrude and George were continually working to improve these new compounds, trying to understand how cancer drugs like 6-MP worked: the process of metabolism in the body, why only cancerous cells were affected, and how they could improve its differential effect. Studying the metabolism of 6-MP was difficult and time consuming. For seven years, they persevered, producing modified versions of 6-MP, in the hope of improving its efficacy. By the late 1950s, other research scientists became interested in one such derivative, the compound now known as azathioprine (Imuran®), which appeared to have an interesting effect on the immune response in transplantation.

Organ transplantation was in its infancy, with the first living related organ transplant of a kidney between twins in 1954. Genetically identical twin transplants do not encounter the same problems with rejection as non-genetically matched transplants. The immune system is designed to respond to any foreign substances, whether these are microorganisms, such as bacteria or viruses, or the 'alien' genetic profile of transplanted organs. Finding a close genetic match between donor and recipient is a key component of a successful organ transplant. The more genetically mismatched the donor organ is, the more quickly and aggressively the recipient's immune response deals with the insult, resulting in rejection of the organ.

In 1958, Robert Schwartz, working with William Dameshek in Boston, found that this particular derivative of 6-MP, azathioprine, prevented an antibody response to the protein albumin, which had been injected into animals in an animal model. Antibodies are proteins produced by B cells (specialised white blood cells) that specifically bind to foreign antigens or targets of the immune response. With Robert Schwartz's help, Gertrude and George set up an immunological screening test, measuring antibody responses, to identify important factors such as: dose, timing of administration and the specificity of the response.

Roy Calne, a young British surgeon, who had heard about Robert Schwartz's work, investigated the effects of azathioprine in kidney-transplant rejection in dogs. Gertrude gave Roy Calne a vial marked #57-322 and he administered the contents to a dog, a collie called Lollipop, who was also given a kidney from an unrelated dog. The drug in the vial was kept secret until 1959, when the success of its trial became evident. To everyone's amazement, the kidney transplant wasn't rejected and Lollipop went on to have a litter of puppies, before dying of natural causes.

In 1962 surgeons first successfully used Imuran® (azathioprine), when performing a kidney transplant between unrelated people. Since then, more than 250,000 kidney transplants have been carried out worldwide, many of them with the help of Imuran®; other organ transplants such as liver, heart and lung have likewise become possible. Immunosuppressive drugs are routinely used to prevent organs being rejected. Other immunosuppressive drugs, for example cyclosporine, have come into use in recent years, but azathioprine remains a mainstay in kidney transplantation.

The immunosuppressive effects of azathioprine have been studied in a wide variety of immunological systems, in particular autoimmune diseases such as autoimmune haemolytic anaemia, systemic lupus, hepatitis and rheumatoid arthritis. Autoimmune diseases occur when the immune system attacks healthy cells in the body, mistakenly recognising them as foreign. The causes are not well understood but involve a combination of genetic and environmental influences. In rheumatoid arthritis, azathioprine suppresses this abnormal response of the immune system, reducing the inflammation that occurs and slowing the damage to the joints caused by the inflammation. The FDA approved azathioprine for use in rheumatoid arthritis in March 1968 and it remains in use today, featuring on the World Health Organization's Model List of Essential Medicines.

In the late 1950s and early 1960s, the lab was working on compounds that would boost the effects of 6-MP. In so doing, Gertrude revealed a purine, allopurinol, which became another success story. Gertrude knew from their metabolic studies on 6-MP that an enzyme called xanthine oxidase was involved in the catabolism (the breakdown of complex molecules to simpler ones) of 6-MP. Enzymes are biological catalysts – substances that increase the rate of chemical reactions without being used up. Xanthine oxidase is responsible not only for the oxidation of 6-MP, but also for the formation of uric acid. Consequently, Gertrude and George found that treatment with allopurinol produced a marked decrease in serum and urinary uric acid, presenting a unique opportunity to develop a treatment for gout.

Gout is one of the most common forms of arthritis, occurring in 2.5 per cent of the adult population in the UK. It often comes on very suddenly, as characteristically red and shiny joints with the attendant pain and swelling, due to the build of urate crystals in the joints. Uric acid is formed from purines during digestion and is present in the blood, circulating as a urate salt. When there is too much urate circulating, urate crystals can accumulate in the joints, causing inflammation. Allopurinol (Zyloprim®) works by targeting the enzyme xanthine oxidase, thereby reducing the production of uric acid.

Like all the drugs they developed, particularly those that are used long term, George and Gertrude had to consider a number of factors. Did the inhibitor have a long enough half-life (the time required for the concentration of a substance in the body to decrease by half) to produce a persistent reduction in uric acid production? By interfering with one part of a biochemical pathway, would other more toxic or equally insoluble products as uric acid result? What were the long-term effects of the inhibitor? Did it interact with other medications? What were the issues if there were other underlying diseases like kidney failure?

All of these possibilities, and more, were carefully examined both in animal trials and clinical trials in man: allopurinol is generally a safe and effective long-term treatment for gout. Untreated, high levels of uric acid in the blood can lead to kidney stones and, potentially, blockages in the kidney. Before the invention of allopurinol, a total of more than ten thousand gout victims died annually from kidney blockage in the USA alone.

In a typical example of Gertrude and George's open minds and desire to share their findings with others, ten years later allopurinol was developed for a quite different therapy, for parasitic diseases like leishmaniasis and Chagas disease (caused by *Trypanosoma cruzi*).

This research relied on the knowledge that parasitic and mammalian enzymes can have quite different specificities and demonstrated the kind of chemotherapeutic selectivity that could be achieved with purine derivatives. Medicine academic and pharmaceutical executive J. Joseph Marr had found that allopurinol inhibits the replication of the parasite *Leishmania donovani* and Gertrude and George were keen to discover the biochemical basis for this unexpected activity. Leishmaniasis is a tropical parasitic disease, caused by infection with the *Leishmania* parasites, which are spread by the bite of infected sand flies. There are several different forms of leishmaniasis in people. The most common forms are cutaneous leishmaniasis, which causes skin sores, and visceral leishmaniasis, which affects several internal organs (usually the spleen and liver, as well as bone marrow).

Like all parasites, Leishmaniae derive nutrients at their hosts' expense and the meal in this case includes the therapeutic drug allopurinol. Leishmaniae lack the ability to synthesise purines but they can use allopurinol present in the mammalian bloodstream. This is taken up by the parasite when it feeds on its human host.

Gertrude and George discovered that the downstream effect of this incorporation of allopurinol into the parasite's metabolism is the inhibition of protein synthesis in Leishmaniae, thereby inhibiting the parasite's replication and inhibiting the development of leishmaniasis. Gertrude encouraged the scientists at Burroughs Wellcome to pursue cures for these debilitating parasitic diseases; even though these parasitic drugs were unlikely to be big money spinners, she was driven by her social conscience and not financial gain.

In 1969, Gertrude decided to return to some research that had fascinated her since 1948 – the anti-viral activity of her first potential cancer therapy, diaminopurine. Infections caused by viruses can be notoriously difficult to treat, for several reasons. Viruses cannot replicate outside a cell and are often hidden from the immune system inside their host cells. In the case of persistent viruses, like the coldsore virus herpes simplex, after the initial infection it lies dormant in nerve endings and only emerges to cause further outbreaks when triggered by, for example, exposure to UV light. The virus hijacks the cell's own replication machinery to make copies of the virus. These viruses burst out of the cell, destroying it, and will attempt to infect many more cells, unless tackled by the immune system. In addition, the immune system has to cope with the ability of many viruses to mutate rapidly and thereby avoid a specific immune response, providing a major challenge to the design of vaccines.

Although most laboratories at the time were focused on vaccine research, Gertrude and George had other ideas. In 1970, the lab moved to North Carolina, where they were joined by new colleagues, including Howard Schaeffer who was head of the Organic Chemistry Department. Along with Schaeffer and John Bauer, a virologist at the Wellcome laboratories in the UK, they spent several years working on a potent compound that was highly active against both the herpes simplex virus and the virus that caused smallpox (vaccinia), but that was less toxic than its predecessor.

Acyclovir (Zovirax®) is a guanosine derivative that acts as an anti-metabolite by inhibiting viral growth. Specifically, it inhibits the action of the enzyme, DNA polymerase, which is involved in the replication of viral DNA. Gertrude was intrigued to find that acyclovir was highly specific to cells infected by herpes viruses only, not other types of viruses, and, importantly, did not damage uninfected cells in the body. As she and her colleagues delved ever deeper into the chemical structures and the pathways involved, the reason became clear. Only a viral specific enzyme, herpes thymidine kinase, could convert the drug into its active form, acyclovir triphosphate.

Acyclovir was introduced to scientists in 1978, at a conference in California. It was a seminal moment for anti-viral research. Thirteen posters were put up in the conference centre lobby, covering all the details of this new drug, everything from its conception to the findings about its mechanism and efficacy. Gertrude's research team of seventy scientists had kept acyclovir under wraps; now it had been revealed, Burroughs Wellcome could patent it and prevent other companies manufacturing it.

Acyclovir is widely used for the treatment of oral and genital herpes, shingles (caused by reactivation of the chicken-pox causing virus varicella zoster) and the most serious medical problem of herpes simplex infections in immunosuppressed individuals. Gertrude later referred to acyclovir as her 'final jewel . . . That such a thing was possible wasn't even imagined up until then.' Zovirax®, the first anti-viral produced, became the world's first billion-dollar drug.

Encouraged by their success with Zovirax®, and in particular its highly selective nature, in the 1980s the Burroughs Wellcome lab joined the search for treatments for a new disease, AIDS (acquired immunodeficiency syndrome), first reported in 1981. AIDS is caused by the human immunodeficiency virus (HIV), which infects T helper lymphocytes, pivotal cells in the immune system. These T cells carry a marker called CD4, which the HIV latches onto and uses to gain entry to the cell, thereby destroying the T cells. The lower a person's $CD4^+$ T cell count, the less able they are to fight infections. Research on AIDS was frowned upon in some quarters. Not only was such a virus viewed as especially difficult to treat, there were those (even in the scientific community) who refused to accept that HIV caused AIDS. Few pharmaceutical laboratories were set up to study viruses and even less viewed this as a profitable endeavour.

Gertrude officially retired in 1983 but remained closely associated with the research as a consultant. Her group collaborated with Dr Samuel Broder at the US National Cancer Institute, who remembers her and Burroughs Wellcome's willingness 'to do things to develop products, and not just to talk. They were willing to exchange information and provide drugs that could be tested. And they were pretty clear that they would make a commitment to try to develop and commercialise a product that looked good.'

Luckily, just as with other compounds they investigated, they had one already waiting to be put to use. Gertrude was not officially involved with this work, but she was credited with providing the inspiration for it. She later modestly commented, 'The only thing I can claim is training people in the methodology: how to delve into how things work and why they don't work and what resistance is and so on. The work was all theirs.' Marty St Clair, a virologist working on treating AIDS with azidothymidine (AZT), did not agree, 'Trudy had everything to do with AZT. Yes, officially she was retired, but she was there working with us and counselling us. She knew exactly what we were doing.'

AZT, one of Gertrude's purine derivatives, was found to inhibit HIV growth in human cell cultures. AZT was the first drug approved for treatment of AIDS. Acting as an inhibitor of the enzyme reverse transcriptase, it stops the reproduction of viral DNA and reduces the amount of virus in the blood (the viral load). AZT was approved by the FDA in 1987 in record time. After animal trials, only one clinical trial was performed on humans, rather than the standard three, and that trial was stopped after nineteen weeks as patients receiving the placebo were dying faster than the treated patients. AZT is not without side-effects but it slows down the growth of HIV significantly. Until 1991 it was the only drug licensed to treat AIDS. Now, AZT (Retrovir®) is one of the standard cocktail of drugs that means HIV infection is no longer an immediate death sentence.

After Gertrude officially retired in 1983 as department head, at the age of sixty-five, she continued as a scientist emeritus and consultant, taking an active part in the research meetings on AIDS and other areas of research, and attending seminars. In addition, she became a research professor of medicine and pharmacology at Duke University, inspiring third-year medical students who were interested in research on tumour biochemistry and pharmacology. Gertrude later commented, 'I think it's a very valuable thing for a doctor to learn how do to research . . . something there isn't time to teach them in medical school. They don't really learn how to approach a problem . . . and I think that a year spent in research is extremely valuable to them.'

Throughout her life, Gertrude had seized the opportunities provided by studying a family of compounds and had thereby generated drugs with a diverse range of therapeutic applications. Her relentless work ethic and inquisitive nature inspired others. She was an important role model for young scientists, who warmed to her friendly and open personality and admired her record of discoveries.

Jonathan Elion, Gertrude's nephew, later said, 'She made herself available to students. While people tell me now that she was an advocate to the advancement of women in science, this actually comes as news to me, as I always thought of her as advocating the advancement of ALL persons in science.' Interacting with students came naturally. Gertrude said, 'In a sense, my career had come full circle from my early days of being a teacher to now sharing my experience in research with the new generation of scientists.' Her legacy of teaching and mentoring continues with awards, such as the Gertrude B. Elion Mentored Medical Student Research Award, which support female medical students who are interested in pursuing health-related research projects.

In 1997, Gertrude said, 'The same thing that inspired me over the years inspires me now. I want to get sick people well. I want to get children involved in science. I want them to have the same sort of excitement and fun that I've had and do something useful with their lives.'

Gertrude's working life had certainly been useful and productive. Her list of patented drugs was in excess of forty-five and she had written over two hundred scientific papers. Her success in the world of pharmacological research is legion. But Gertrude hadn't always felt her scientific career would take off. Although she had registered for a PhD in her late twenties, the demands of her newly acquired job in George Hitching's lab meant that she had reluctantly curtailed her studies.

The validity of her decision not to lose the job she loved in pursuit of a PhD was borne out much later in her career as she received no less than twenty-five honorary degrees. And Gertrude's contribution to science was recognised with numerous awards. For example, in 1968, she was given the Garvan Medal from the American Chemical Society, the only woman to receive this prestigious award until 1980 and one of the first signs of approval from the scientific community. Gertrude was reportedly so affected by this recognition that she was moved to tears.

Twenty years later, on 17 October 1988, Gertrude was washing in her bathroom, at 6.30 a.m., when the phone rang. It was a reporter, ringing to tell her that she had won the Nobel Prize. Gertrude thought he was joking until he mentioned who she was sharing it with: Sir James W. Black and George H. Hitchings. The three were awarded the Nobel Prize in Physiology or Medicine 'for their discoveries of important principles for drug treatment'. Sir James Black had discovered two classes of drugs: beta-blockers for high blood pressure and heart disease, and H-2 antagonists for ulcers.

In typical fashion, Gertrude, now aged seventy, refused to be overawed by the prize-giving ceremony. As the only female recipient of a Nobel Prize that year, she stood out in her blue chiffon gown when everyone else present was in black and white. She brought eleven members of family with her: her nephews and niece and their families, including four children under the age of five. Gertrude insisted the children be allowed to attend the formal banquet, saying, 'I'm not going to bring them all the way to Sweden and then have them spend the evening in a hotel room. You put them at a separate table where they can see their parents and their parents can see them and they'll be fine.' And sure enough, they were, behaving beautifully and entertaining all present.

Gertrude's informal approach belied the seriousness of her work and the rarity of pharmaceutical industry employees and women receiving the Nobel Prize. Asked about which of the drugs she had discovered were the most important to her, Gertrude replied, 'It's like being asked to discriminate amongst your children.' She found the 'birth' of every drug equally exciting and important and recognised that, 'each one in its own time was kind of a revolutionary drug'.

As she neared the end of her life, Gertrude could look back with satisfaction. She and George Hitchings had revolutionised the process of drug discovery, and the drugs they developed were instrumental in saving and improving millions of people's lives. In the 1950s, the prospects were bleak for children with acute lymphoblastic leukaemia. Gertrude's and George's 6-mercaptopurine is one of the great success stories in the fight against cancer. Acyclovir was the world's first anti-viral. Their AIDS drug AZT continues to be in widespread use and is one of the most important long-term results of the programme of rational drug design that George Hitchings began in 1942. Imuran® remains an essential drug in transplantation. 'Now, I didn't start to make a compound that would do that,' Gertrude later pointed out. 'But, if you listen and keep your mind open, this is what can happen. This was the story of our lives.'

In her Nobel Prize acceptance speech, Gertrude concluded by saying, 'I hope that I have successfully conveyed our philosophy that chemotherapeutic agents are not only ends in themselves but also serve as tools for unlocking doors and probing Nature's mysteries. This approach has served us well and had led into many new areas of medical research. Selectivity remains our aim and understanding its basis our guide to the future.'

If a compound altered a step in the way nucleic acid was synthesised, Gertrude and George used it to explore that step and its downstream effects. This leverage technique was one of the key elements of their drug design method and is still used in many of the drugs developed by Burroughs Wellcome's successor GlaxoSmithKline (GSK). For example, nelarabine has a role in treating certain rare forms of leukaemia and lymphoma when other treatment options have been exhausted. The journey to its licence in the USA in 2005 started in the 1980s when Dr Joanne Kurtzberg, a paediatric oncologist, was trying to develop a new drug for paediatric patients. 'She [Gertrude] gave me two little glass vials, I still remember, with black tops.' In one of these vials was a precursor of nelarabine. Nelarabine is no panacea. Like many cancer drugs, it is toxic, but it can provide a window of opportunity to perform a bone-marrow transplant.

Starting with 6-MP, Gertrude had always tried to find drugs to help cancer patients who desperately needed additional methods of treatment. An oncologist who worked with GSK, Dr Neil Spector, remarked, 'Her words still ring out to me, "Neil, you have to keep an eye on the patients. If you do that the company will do fine."'

With the help of Gertrude and George's battery of drugs, Burroughs Wellcome did more than fine and, importantly for medical research, so did the Wellcome Trust. When Sir Henry Wellcome died in 1936, he left two legacies: the pharmaceutical company he had co-founded with Burroughs and his medical research charity, the Wellcome Trust, which was the sole shareholder of the company. During the Second World War, the focus on supporting the UK war efforts had left the pharmaceutical company nearly bankrupt. The company's fortune had a direct impact on the Wellcome Trust, which relied on dividends to fulfil its charitable objectives of supporting research to improve human health.

Burroughs Wellcome, which was run in the USA as an independent offshoot of the parent company, was so profitable through the war years that it ensured the survival, and thereafter decades of growth, for the UK parent company. This was in no small part due to the efforts of Gertrude Elion and George Hitchings, and their landmark and highly profitable drug discoveries. The pharmaceutical company became so valuable that, after the merger with Glaxo Plc in 1985, stock market flotation in 1986 and diversification of its assets over the next fifteen years, the Wellcome Trust is now the largest medical charity in the world.

After her official retirement from Burroughs Wellcome in 1983, and alongside her mentoring and advisory roles, Gertrude went to work for the World Health Organization. She had developed drugs for leishmaniasis, Chagas disease and malaria and was well-placed to serve these causes. Gertrude worked on several committees, including the tropical disease research division. Her dedication to the WHO ensured that her work spread to countries less developed than the USA.

Gertrude felt that her biggest legacy was the drugs she helped develop. When asked how it felt to receive the Nobel Prize, she replied, 'It's very nice but that's not what it's all about. I'm not belittling the award. The prize has done a lot for me, but if it hadn't happened, it wouldn't have made that much difference … When you meet someone who has lived for twenty-five years with a kidney graft, there's your reward.' She said, 'What greater joy can you have than to know what an impact your work has on people's lives? We get letters from people all the time – from children which are living with leukaemia. You can't beat the feeling you get from these children.' One of her many letters of gratitude said, 'After a very severe case of shingles, Zovirax saved my eyesight. If you ever feel unappreciated for any reason, please take out this letter and reread it.'

These letters brought home to Gertrude how much her research affected the lives of real people, including those closest to her. Cancer had made a huge impact, not only in the success of the cancer drugs she had developed, but in her personal life – firstly, the loss of her grandfather to stomach cancer in 1933 and finally, her fourth loss, her mother to cervical cancer in 1956. Gertrude was only thirty-eight when her mother died and later realised how much she would have relished and appreciated Gertrude's drug discoveries.

Gertrude's successful career was due partly to her workaholic nature, but she enjoyed life too. In retirement, she found more time for her hobbies: photography, trips to the opera and time spent with her brother's family – one niece and three nephews. With her good friend and neighbour, Cora Himadi, she travelled, visiting Africa, Asia, Europe and South America. Her two-storey townhouse, which she had lived in ever since 1970, was filled with the art she collected.

Gertrude's life ended abruptly after she collapsed while out on a walk. She died aged eighty-one a short time later, in a North Carolina hospital, at midnight on 21 February 1999. Gertrude's death was a shock to all those who knew and loved her. After she died, her nephew Jonathan was sorting through her mail and found a number of letters from appreciative patients and grateful colleagues. One was from a young girl who, as Johnathan commented, 'talked excitedly about a school project . . . she had researched scientists on the internet and had selected Trudy as her heroine'.

Gertrude may have regarded herself as an unlikely candidate for the status of heroine, but she was fearless in her pursuit of some outstanding drugs that alleviate suffering worldwide. She was often asked to give advice to women going into science. Her reply sums up her pursuit of science: 'I have no mysterious secret to impart. The most important advice is to choose the field that makes you happiest. There is nothing better than loving your work. Second, set a goal for yourself. Even if it is an "impossible dream", each step towards it gives a feeling of accomplishment. Finally, be persistent. Don't let yourself be discouraged by others, and believe in yourself.'

CHAPTER FIVE

Dorothy Hodgkin
(1910–94)

In October 1964, the British press had a field day. The *Daily Mail* headline read 'Oxford housewife wins Nobel', while the *Telegraph* declared 'British woman wins Nobel Prize – £18,750 prize to mother of three'. Dorothy Hodgkin, aged fifty-four, had been awarded the Nobel Prize in Chemistry for revealing the structure of penicillin and vitamin B12. She is the first and only British woman to win a Nobel Prize in science. Her ability to 'see' such molecules transformed the field of X-ray crystallography, helping biologists to understand how proteins function and guiding therapeutic medicine.

Dorothy's work was unique, not only for its technical brilliance and medical importance but also for her use, at every step, of increasingly complex computers and her ability to forge international collaborations. Dorothy's personal and scientific story is one of gentle persistence in the face of adversity, combined with a lifelong interest in people and their needs.

Dorothy Mary Crowfoot was born on 12 May 1910, the eldest of a family of four girls. She spent her early years with the exotic backdrop of the Pyramids and living the comfortable, sociable life of an expat family in Cairo. Her father, John Crowfoot, had moved to Egypt as a civil servant in 1901 after studying classics at Oxford. By the time Dorothy was six, he had been promoted to be the director of education in the Sudan and the family moved to Khartoum. The dinner-party circuit was just one facet of the family's existence; both parents were enquiring characters who encouraged learning and intellectual pursuits, particularly in the field of archaeology, and were firm believers in selflessness and service – traits that Dorothy inherited.

Dorothy's mother came from a background that promoted genteel activities for women but left her frustrated intellectually. Grace Mary Hood (known as Molly) was the eldest of six children, including four boys, who lived at Nettleham Hall in Lincolnshire. While the boys made careers in, for example, the army or the navy, Molly, much to her frustration, was educated at home and did not fulfil her early ambitions to study medicine.

The first four years of Dorothy's life were spent with her sisters, Joan and Elisabeth, in Cairo, returning to England in the summer months to escape the relentless heat. In 1914, Dorothy's world, and that of countless others, was turned upside down. The year marked both the outbreak of the First World War and the last time Dorothy and her sisters would live with their parents for more than a few months at a time. For the duration of the war, they led a quiet existence with their much-loved nanny, Katie Stephens, and their paternal grandparents in the southern seaside town of Worthing. As the eldest girl, Dorothy subsequently assumed some responsibility for her sisters when Katie married and went to live in Australia. The seeds of Dorothy's resilience and quiet independence were sown.

While John Crowfoot carried on his work in Khartoum, Molly Crowfoot returned to the UK for a while, with Dorothy's new sister Diana, born in 1918. Dorothy was brimming with energy and a passion for learning, so Molly set about teaching her children herself. With little formal education, her methods were unorthodox but successful. Dorothy and her sisters loved the geography lessons, drawing maps on the mud floor of the greenhouse, designing and illustrating their own history books and collecting nature specimens on their walks.

As John Crowfoot approached retirement in 1920, and Dorothy her tenth birthday, the family planned for the next few years. Molly was keen to both support her husband in their African life together and set up a family home in the UK. This was a ramshackle house, appropriately called the Old House, in Geldeston, Norfolk, where Dorothy started her forays into the world of chemistry. She was inspired by a teacher who understood the importance of practical demonstrations. Dorothy watched entranced as the class made the unforgettable, iridescent blue crystals of copper sulfate. At eleven years old, and encouraged by her free-spirited and tolerant parents, she used her pocket money to buy materials from the local chemist and started experimenting in the attic. With little regard for health and safety, Dorothy entertained herself and her sisters with the colourful, and often explosive, world of chemistry.

A year later, Molly returned to Khartoum, spending half the year there, while the children lived with relatives or friends. Dorothy launched herself into life at the Sir John Leman School in Beccles, Suffolk. In 1921, the school had 130 pupils, with girls in the majority, as many boys were sent away to public school. Dorothy found her eclectic education to date meant that she was more than up to speed with the likes of English and history but less so with maths and chemistry, which were both taught by their needlework teacher, Miss Christine Deeley.

Every opportunity to probe the world of chemistry was taken up, even to the extent of utilising a nosebleed. Dorothy thought, 'it was a pity all this good blood should go to waste so I collected it in a test tube and used it to isolate [the blood pigment] haematoporphyrin'. Molly encouraged her interest, buying her the published version of two chemistry lecture series for children, given at the Royal Institution by Sir William Bragg, in 1923 and 1925. Bragg and his son Lawrence had pioneered the use of X-rays to study the structure of materials. And her cousin, Charles Harington, who isolated the hormone thyroxine, recommended the first 1921 edition of a biochemistry textbook, *Fundamentals of Biochemistry* by T. R. Parsons. These influences combined to set Dorothy on her career path.

To achieve her desire to study chemistry, Dorothy had to work hard. At fifteen, she was already in a class with students who were on average eighteen months older than her. In March 1927, at aged seventeen, the school leaving certificate exams loomed. These would allow her to apply to university. Progress wasn't always easy. She was a perfectionist at heart and wanted to understand the intricacies of all she studied. When there were tears of frustration over a maths problem, her mother tried to reassure her that the answer was correct. Dorothy retorted, 'Of course it's right, but I can't see why!'

Dorothy passed with flying colours, achieving the highest marks for a girl in the local school leaving certificate papers that year. With subsequent tuition from a range of tutors, Dorothy's drive to succeed was rewarded with a place to read chemistry at Somerville College,

Oxford. In the summer of 1928, before she started university, Dorothy joined her parents on an archaeological dig in Jordon. There she worked with her father reconstructing fifth- and sixth-century pavement mosaics, a painstaking and difficult task. Dorothy was fascinated by their intricate patterns and later referred back to her scale drawings of these when she was studying the 3D structure of crystals. Her ability to visualise the often hidden atomic structures became a valuable skill in her interpretation of X-ray photographs. As her son Luke later said, her mosaic work, 'meant that Dorothy knew what was involved in piecing things together in detail'.

Oxford did not come as a complete shock to Dorothy. She was well used to the discipline of independent study and her allowance of £200 a year precluded any immediate financial worries. The male-dominated environment presented more of a challenge. Women had been studying at Oxford for fifty years but only received a degree for their efforts from 1920 onwards. In 1928, women were still excluded from certain societies, some lecturers would evict women from the lecture theatre and there were prohibitive rules governing socialising between the sexes.

The upside to the chauvinistic nature of that time was the dedication and passionate belief in women's education in the women's colleges. Champions of the cause put great value on the highest level of scholarship, with a stimulating and intellectual environment that suited Dorothy. In her first year at Oxford, Dorothy entirely immersed herself in her work. By the second year, some of the other attractions of Oxford had penetrated her consciousness and she was now a member of several societies.

Dorothy had also made friends, including Elizabeth ('Betty') Murray, an outgoing and fun-loving history student whom Dorothy met at the Archaeological Society and the Labour Club meetings. Betty took Dorothy under her wing, involving her in social events and regular walks, and providing moral and practical support. Betty shared Dorothy's joy at her laboratory successes but was horrified by the hours she put in, commenting, 'I think she looks very thin and not too well, but of course there is no stopping her working.'

Dorothy's ability to be single-minded bore fruit in her last year at Oxford. After the final exams, the fourth year was spent entirely on research – an enduring and distinctive feature of chemistry at Oxford since 1916. Dorothy found that she was most fulfilled by her time in the laboratory and decided that she would like to discover the structures of complex molecules using X-ray crystallography. She was driven by the conviction that the X-ray image was the best basis for understanding the chemistry and function of molecules.

This understanding can only come from knowing how atoms fit together at the tiniest level of detail. For example, diamond, graphite and graphene (discovered in 2004) are all composed of just carbon atoms, but diamond is rock hard, graphite flows onto the page from a pencil and graphene is one of the lightest, thinnest but strongest materials ever discovered. The unique structural arrangements of their carbon atoms lie at the heart of their individual properties. Biological molecules, such as hormones, lock onto their receptors on the surface of cells with a specific shape that facilitates that interaction. The enzymes that catalyse chemical reactions can only do so when they have the correct conformation. Understanding the shape of molecules, such as proteins, is vital to not only define their function but also for the design of new and improved medicines.

Chemists often know what atoms make up a molecule and in what proportions. But in the 1930s, and often nowadays, their arrangement, the way they fit together, was frequently unknown, particularly for large proteins, which have hundreds or thousands of atoms. The technique of X-ray crystallography allows scientists to work out how the atoms fit together in space, so as to create the molecule's shape.

By making pure substances, scientists such as Dorothy were able to crystallise molecules. She knew that the flat surfaces and precise angles of crystals reflected the regular arrangements of the atoms inside them. X-ray beams were fired through the crystal and the diffracted X-rays were picked up on a photographic film. A similar, but more automated, process occurs in modern X-ray crystallography machines. The X-rays are diffracted by the molecule's atoms in different directions and in such a way that some of the waves reinforce each other. When the diffracted X-rays hit the photographic film, there is a pattern of spots with varying intensity.

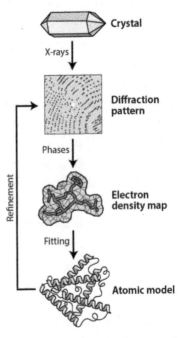

X-ray crystallography: scattering of X-rays by the atoms of a crystal produces a diffraction pattern that gives structural information on the crystal – an electron density map – that allows scientists to identify the crystal and produce an atomic model

In Dorothy's time, the analysis of these X-ray diffraction patterns was initially done by eye, comparing with a reference pattern, an extremely tedious and time-consuming process. Complex mathematical analyses were also used. Finally, a ball-and-stick model was created, allowing Dorothy to physically handle and turn around a representation of the molecule on her desk.

In September 1931, aged twenty-one, Dorothy started her undergraduate research project. This involved the first significant use of X-ray crystallography at Oxford, working on compounds called thallium dialkyl halides – intermediate structures between the simple molecules that had been analysed to date and the much more complex proteins that were still to be analysed.

The X-ray equipment was housed in a vast room on the first floor of the University Museum. Much of what Dorothy was using was primitive by today's standards and, rather like her childhood attic experiments in Geldeston, did not conform to any currently acceptable health-and-safety rules. Her vigilant friend Betty reported that she was worried 'when she doesn't come in as the machine she works on is a dangerous one. She did give herself a bad electric shock one day this week which if the current had been full on would have killed her.'

Geometrical and mathematical analysis of the pattern of spots started to yield the first crystal structures. The maths involved was challenging and Dorothy worked long hours, often late at night, but she persevered and eventually her hard work paid off. In 1932, aged twenty-two, she was awarded a first-class degree, following in the footsteps of just two other women. She was now in a strong position to join John Bernal's laboratory in Cambridge as his PhD student.

John Desmond Bernal was an enthusiast who was incredibly knowledgeable and had an opinion about a wide range of subjects, earning him the nickname 'Sage'. He was also an ardent socialist, part of a strong left-wing group in the Cambridge research laboratories and founder of the Cambridge Scientists' Anti-War Group. John liked women, to the detriment of his marriage but to the benefit of the laboratory and Dorothy.

John Bernal influenced Dorothy on a number of levels. As a scientist and a socialist, he was a man who saw science as both a form of knowledge and a force for change. British science in the 1930s was poorly funded and disorganised and John thought the scientific disciplines should be more integrated and work together to translate their findings for society's benefit. Although his ideals were born out of a communist Utopian vision, he was ahead of his time in recognising the need to reposition science centrally in society.

Dorothy followed John's views with interest but was not actively involved in propagating them. She had, however, fallen for his magnetic personality; he, in turn, was equally captivated. Dorothy conducted their relationship in her characteristically unobtrusive fashion and, perhaps bearing in mind that she was his research student, he was equally circumspect. The relationship was not to last, but they did remain friends and in close personal and professional contact for the rest of their lives.

Just as her research was taking off in Cambridge, the first symptoms of rheumatoid arthritis occurred. Dorothy was developing pains in the joints of her hands and had been persuaded by her parents to have them investigated. Despite the attention of a Harley Street doctor, her condition went undiagnosed and, after a short break, Dorothy continued her work.

John Bernal was at the forefront of understanding the structure of biological molecules using X-ray crystallography. Sterols, like cholesterol, formed the basis of Dorothy's PhD thesis and her output was prodigious. Publishing scientific papers is one of the hallmarks of a successful scientist and, by the mid-1930s, she was co-author of a number of papers, establishing her early prominence in the field. John was one of the examiners for her 300-page thesis. He described it as 'the first comprehensive attempt at a joint crystallographic and chemical study of a group of substances of great intrinsic interest as well as critical biological importance'. He may have been somewhat biased but the scientific world was completely in agreement.

In the 1930s, very little was known about the structure of many molecules, including complex proteins and DNA. It's now understood that, at the simplest level, each protein contains chains of amino acids, of which there are at least twenty different types in the human body. The unique sequence of amino acids that makes up a protein, or polypeptide chain, is called the primary structure. The secondary structure is the way in which this chain then folds, in pleats or coils. Finally, the tertiary or 3D structure results from further interactions between, and shaping of, these folds.

Dorothy was one of a number of prominent scientists pushing the frontiers of knowledge forward. Working with proteins was not easy. Proteins do not naturally lend themselves to crystal formation and their 3D structure was proving elusive. One such protein was pepsin, the main digestive enzyme in the stomach, which itself breaks down proteins into polypeptides. John and Dorothy discovered that the key factor was keeping the pepsin crystals wet. On 26 May 1934, they published their research in a top-tier scientific journal, *Nature*. The paper was entitled simply: 'X-ray photographs of crystalline pepsin'.

By combining new and complex mathematical techniques, like Fourier's transform and Patterson maps, with isomorphous replacement, the field of protein X-ray crystallography was advancing. Niels Bohr, the Nobel Prize-winning physicist, had developed the idea, in 1913, that an atom is a small positively charged molecule surrounded by orbiting negatively-charged electrons. Isomorphous replacement involves the addition of electron-dense or 'heavy' atoms to a protein, without disturbing its structure, but making the X-ray diffraction patterns of complex molecules easier to interpret. At every stage of her research career, Dorothy wasted no time in adopting the newest methods.

Dorothy was now settled in Cambridge and enjoying her life in John Bernal's laboratory, but Oxford was keen to have her back. In 1934, aged twenty-four, she accepted a research fellowship at Somerville College, with an unusually small teaching responsibility, allowing her to get on with research.

Understanding the structure of molecules was not the only focus and attraction of her life in Oxford. Dorothy had been welcomed back with open arms and was soon established socially too, enjoying the regular teas and dinners that were a feature of college life. Social distractions were to reach their peak in spring 1937. On a visit to London, Dorothy stayed with Margery Fry, her Somerville college principal, where she met Margery's cousin Thomas Hodgkin. After gaining a first-class degree in classics from Balliol College, Oxford, and spending a year in Palestine, Thomas was now looking for a job back home. Dorothy immediately found him interesting and charismatic; it was both a love match and a meeting of minds.

For once, Dorothy was distracted from her work and wrote to Thomas that she was 'a little troubled by your urgency'. Not for long, it seems, as their engagement was announced and Dorothy was overjoyed, writing that 'I think whatever happens this will be one of the happiest days of my life.' The couple were married on 16 December 1937.

Thomas now had job as a history lecturer at the Friends Voluntary Service for Unemployed Miners in Cumberland. Dorothy was also desperately keen to continue her work. Before she could do so, she had to deal with the marriage bar, the practice restricting women from working once married. From the mid-1920s, it was common for women to give up work when they got married. The marriage bar was finally abolished in 1944 in teaching and in 1946 in the civil service, but not before a generation of married women had been forced to give up jobs they had trained for and were good at.

In 1937, aged twenty-seven, Dorothy had a permanent fellowship and she duly wrote to offer her resignation, as was expected. Thankfully, Oxford was sufficiently enlightened to take Dorothy back immediately. Dorothy's growing reputation must have worked in her favour but other women were not so fortunate. In the 1930s, only 7.5 per cent of higher professionals, for example doctors and lawyers, were women and only 12 per cent of married women worked outside the home.

In sharp contrast (according to the 2013 UK government's Office for National Statistics report), at least 60 per cent of women with primary school-aged children are in paid employment today. This rises significantly as children get older. Dorothy's work was vital to her for the stimulus and satisfaction it provided, but its permanent and relatively well-paid nature may also have driven her desire to continue working after her marriage.

The early years of the Dorothy and Thomas's marriage were conducted at a distance, by mutual agreement, as Thomas continued his teaching in Cumberland and Dorothy her research in Oxford. They corresponded daily and Thomas came to Oxford at the weekends. The start of the Second World War was fast approaching and Somerville was a hotbed of thinkers and political movers and shakers. The anti-war logician and philosopher Sir Bertrand Russell and Lord and Lady Longford (then known as Frank and Elizabeth Pakenham) were among the guests at Oxford dinner parties. Left-wing political feeling was prevalent and Dorothy continued to be a regular attender of Labour Club meetings.

Dorothy and Thomas had three children: Luke, Elizabeth and Toby, born in 1938, 1941 and 1946 respectively. At the age of twenty-eight, Dorothy announced the news of her first pregnancy with some relief. A group of physicists at Oxford had raised concerns about her constant exposure to the potentially damaging effects of X-rays and she had felt the need to have a through medical examination before she was married. Luckily, she had been given the all-clear.

Now, Dorothy had another hurdle to overcome. After her success negotiating the marriage bar, she had to tackle Somerville's attitude to motherhood and working. Thankfully, the principal Helen Darbishire was on her side and Dorothy described her as 'very sweet and reasonable'. Dorothy became the first woman at Oxford to receive paid maternity leave; statutory maternity leave was not routinely granted in the UK until 1975.

Luke Howard Hodgkin was born on 20 December 1938. After a

challenging maternity leave in which she developed a breast abscess requiring surgery and Thomas had a car accident, Dorothy geared herself up for work again. Then, the first acute attack of rheumatoid arthritis occurred. At aged only twenty-eight, her joints seized up and everyday tasks like walking upstairs or getting dressed were intensely painful and difficult. Her Harley Street doctor referred Dorothy to Dr Charles Buckley, who ran a specialist clinic in the spa town of Buxton in Derbyshire. He prescribed a mixture of a month's recuperation at the spa, with mudpacks and paraffin-wax hot baths, combined with gold injections. Buckley was realistic about the relapsing nature of rheumatoid arthritis, which meant Dorothy was, in some ways, prepared for her future attacks.

Dorothy's remarkable levels of focus, to the exclusion of her pain and progressive disability, stood her in good stead as she continued to strive in her work and for her family. Although it was unusual at the time to continue working as a mother, it was still quite common in the UK in the 1930s to employ domestic help. A live-in nursery nurse and a part-time cook and cleaner were taken on, greatly facilitating Dorothy's return to work. Here, the effects of her illness were more apparent, when she found it impossible to switch on the X-ray equipment in the laboratory. Her X-ray technician, Frank Welch, made her a long lever and the work continued.

As the Second World War broke out, Dorothy carefully planned her working day so she could spend time with Luke and they had a succession of helpers – often refugees fleeing the war in Europe. Wartime rationing did not prove too arduous for Dorothy. Friends from her Oxford days had commented on the austere nature of her rooms there. Now, she grew her own vegetables and, amazingly, found the time to make clothes for Luke.

In 1941, when Dorothy was thirty-one, her second child Prudence Elizabeth, known as Elizabeth, was born. Second-time around, things were easier and Elizabeth was a contented baby, feeding and growing well, as Dorothy wrote to Thomas, 'Lisbeth in my arms. She weighed 9lbs 4oz this morning and has gained an average of 7½oz per week in the first eight weeks of her life.' Dorothy returned to her research a few months after Elizabeth's birth, reporting a 'triumphal entry to College yesterday' and how she 'naturally bragged about my daughter for the next ten minutes after which we reluctantly began the meeting'.

Toby Hodgkin was born in May 1946, completing the family. His arrival was swiftly followed by another triumph. Dorothy was elected as a fellow of the Royal Society, following only two other women: the Cambridge biochemist Marjorie Stephenson and Kathleen Lonsdale. Kathleen had been a protégé of Sir William Bragg and, like Dorothy, she was married with three children. The Royal Society is a self-governing fellowship made up of the most eminent scientists across the UK and the Commonwealth. Fellows are elected for life through a peer-review process based on their research achievements. In the week she was elected, Dorothy felt 'kind of dotty all the time'. Her elation was matched by some of the comments she received, including those of Alan Hodgkin (Thomas's cousin, a future Nobel Prize-winner in Physiology) who said, 'I think it is magnificent that you manage to combine looking after a family with research, not to mention University teaching. I complain when I have to wash the dishes of an evening.'

By the end of the Second World War, Thomas had also been successful in his work. After years of teaching in adult education in Cumberland, he had a new teaching job in Oxford and the family were reunited. In 1948, Thomas became immersed in a love affair with Africa. He wrote extensively about African history and became passionately involved in the move to African self-rule. In 1951, he acted as an advisor to Kwame Nkrumah, who became the first president of an independent Ghana. In 1962, Thomas returned there for three years to head the new Institute of African studies at the University of Ghana.

At home, although Thomas's charismatic personality took

centre stage, he was extremely supportive of Dorothy's career. But money was tight and Dorothy quickly became the major breadwinner. In 1951, the family had moved from a flat in Bradmore Road to a house on the outskirts of Oxford. Less convenient for Dorothy's work, Powder Hill did have a spacious garden for their expanding family.

Combining her scientific research successfully with family life was a priority for Dorothy. In her opening speech as president of the International Union of Crystallography in Kyoto in 1972, she quoted Kathleen Lonsdale's thoughts on working and bringing up a family, apparently recognising some of herself and her working practices: 'She must be able to do with very little sleep, because her working week will be at least twice as long as the average trade unionist. She must go against her early training and not care if she is regarded as a little peculiar. She must be willing to accept additional responsibility, even if she feels that she has more than enough. But above all, she must learn to concentrate in every available moment and not require ideal conditions in which to do so.'

Dorothy's ability to do all these things was to yield significant results in the field of X-ray crystallography, succeeding in solving the structures of some of the most complex biological molecules.

In the early 1940s, Dorothy worked on penicillin, the drug that heralded the widespread use of antibiotics and revolutionised the field of therapeutic medicine. In 1928, at St Mary's hospital in London, Alexander Fleming had noticed that mould had developed accidentally on a set of culture dishes being used to grow staphylococci bacteria. The mould had created a bacteria-free circle around itself. Fleming named the active substance in the mould penicillin.

In Oxford, the Australian Howard Florey and Ernst Chain, a refugee from Nazi Germany, were investigating the effects of penicillin *in vivo*. In a pivotal experiment, they showed that mice injected with a lethal dose of the bacteria streptococcus only survived if they were also given an extract containing penicillin. Penicillin looked like a wonder drug but large-scale production from the fermentation process was time consuming and inefficient; producing a pure synthetic form of penicillin was impossible until the structure was known.

Dorothy was joined in the laboratory by Barbara Low, who found that growing penicillin crystals was not easy. Apart from their tendency to form a glue-like mess, the precise chemical formula was also unclear. In July 1943, the discovery of a sulfur molecule completed the list of known atoms in penicillin – carbon, hydrogen, nitrogen, oxygen and sulfur – but the formula, the number of atoms of each element, was still being debated.

Penicillin was proving to be a highly political molecule too, with the Americans also in hot pursuit of its structure. Scientists in the USA had succeeded in obtaining crystals of a sodium salt of penicil-lin – a simpler form of the molecule than the one Dorothy had been working on and therefore it was potentially easier to deduce its struc-ture. Dorothy got in contact with Sir Henry Dale, director of the Royal Institution (since the death of Sir William Bragg in 1942), to enlist his help in obtaining a sample from the American company Merck. In February 1944, a military aircraft arrived in the UK with 10 milligrams of the penicillin on board. Dale's colleague and Dorothy's fellow X-ray crystallographer, Kathleen Lonsdale, deliv-ered it to Dorothy in Oxford.

Using the crystals Kathleen had brought, along with other salts of penicillin, Dorothy and Barbara Low yielded their first set of data, with the help of Charles Bunn at ICI's Alkali Division. He was a pioneer in the use of the 'fly's eye' method, where an optical version of the X-ray diffraction pattern was compared to the actual version obtained from a crystal. This approach massively speeded up the analysis. The project also generated so much data that Dorothy used an early type of computer, called a Hollerith punched card machine. As the structure emerged in 1945, they discovered that the penicillin molecule did not have an elongated form, as first suspected, but rather existed in a curled-up state. Perhaps most importantly, Dorothy's unambiguous demonstration of the 3D location of penicillin's atoms made X-ray crystallography the definitive technique for the structural analysis of biological molecules.

Dorothy's contribution was not nationally or internationally revealed then, as industrial secrecy was still of some concern. The intense intellectual satisfaction, which made her 'extremely cheerful', eventually resulted in a full account of the penicillin story, entitled *The Chemistry of Penicillin* and published in 1949. At thirty-nine atoms, penicillin is a small but powerful molecule, and heralded an era of new antibiotics and the end of the tyranny of bacterial infections.

As Dorothy's successes at work continued, her growing reputation and international standing were still not fully recognised by the Oxford hierarchy. Eventually, in May 1945, aged thirty-five, Dorothy became a university employee for the first time as a demonstrator in chemical crystallography, after ten years of extensive research activity. This had the immediate effect of improving the family's financial position but unfortunately did not come with improved working conditions. She was to remain in the cramped and unsuitable conditions of the University Museum for another twelve years.

Dorothy's position as a world-class crystallographer was established by the early 1950s and the main focus of her laboratory was vitamin B12. This is an essential vitamin, which plays a vital role in the creation of healthy red blood cells and in the function of the central nervous system. Deficiencies in vitamin B12 can be cured by taking a synthetic supplement. In 1926, an American scientist, George Minot, demonstrated that the autoimmune disease pernicious anaemia could be cured by an extract from the liver. The extract was subsequently shown to contain vitamin B12 and the first crystals were obtained in 1948 by chemists at Merck in the USA.

Vitamin B12 contains nearly two hundred atoms, not very large by the standards of most biological molecules, but by far the largest to be analysed at the time. Two factors contributed to Dorothy's success with vitamin B12. Although the precise formula of vitamin B12 was still unknown, it was known to contain a cobalt atom, heavy enough to show up on the Patterson maps, now being used in X-ray crystallography analysis. By working on a molecule that already contained a heavy atom, Dorothy had an easier job than the Cambridge-based scientists, Max Perutz and John Kendrew, who were starting to analyse protein molecules using isomorphous replacement to artificially add heavy atoms.

The American scientists at Merck and their British counterparts in the UK were in intense competition to solve the structure, but Dorothy's wealth of experience in interpreting the data came into play. She could clearly see the X-ray diffraction pattern that corresponded to the heavy cobalt atoms but there were also vague images of something else. A young PhD student, David Phillips, was present at her talk to the second International Congress of Crystallography in Stockholm in 1951 and later commented, 'Dorothy said ... "It looks like a collection of pyrrole rings", and everyone looked at her blankly because they couldn't see anything at all. That was really my first introduction to Dorothy and her imaginative, inspirational way of going about the subject – her dependence on the interpretation of electron density maps backed up by chemical understanding that was very rarely brought into the foreground.'

Another factor that contributed to Dorothy's success was the early model of the Hollerith machine. This first automatic data processing system had been developed by Herman Hollerith, a USA statistician who first used the machine to count the American census of 1890. It wasn't until a more powerful computer was available, however, that significant progress was made, with the help of Ken Trueblood, an American crystallographer. He was working at the University of California in Los Angeles with the powerful National Bureau of Standards Western Automatic Computer or SWAC. Even with SWAC, progress wasn't easy. Ken Trueblood made a mistake in calculating the position of one of the atoms. Downtrodden, he reported, 'the blow, coming on top of only about 14 hours sleep in the last three nights, is just too much'.

As vitamin B12's structure was slowly revealing itself, the contribution of Dorothy and her colleagues was recognised. The American biochemist Linus Pauling wrote, 'I am writing to congratulate you on the wonderful job you have done with Vitamin B12. I find it hard to believe, although very satisfying, that the methods of X- ray crystallography can be used so effectively on such a complex molecule.' His approval was important because he had produced a set of rules as to how atoms form crystals, and now here was the evidence.

In 1954, Dorothy published a partial structure and two years later a tentative structure of the whole vitamin B12, both in *Nature*. Its structure revealed the existence of a previously unsuspected chemical grouping, the corrin nucleus. Throughout the eight-year process, the significance of the various laboratories involved, both in the USA and the UK, was the subject of some discussion. Dorothy constantly tried to be fair and, in both *Nature* papers, she took care to acknowledge, and name as authors, all the scientists involved. Bragg later described Dorothy's achievements in solving the structure of vitamin B12 as 'breaking the sound barrier' in the field. Now the full structure of this life-saving vitamin was known, it could be manufactured on a large scale.

In 1960, aged fifty, Dorothy was eventually appointed to a personal chair, the Wolfson Royal Society Research Professor. This gave her security with a decent salary, money to spend on research and freedom from teaching responsibilities. On hearing the news, Dorothy tried to speak to Thomas, who was in Ghana. There was some difficulty getting hold of him and the operator had to keep the line open, at Dorothy's expense. When the happy news had been relayed, the operator's feelings on the matter were quite clear: 'My word, what a fuss about a chair! Now that you to are to have it, I do hope you find it most comfortable.'

In the early 1960s, Dorothy returned to the molecule that had challenged her for nearly four decades. The hormone insulin is produced by the pancreas to control the amount of sugar in the blood. In type 1 diabetes, an autoimmune disease, the immune system attacks the cells in the pancreas, destroying them and reducing the production of insulin. Injection of insulin is the only treatment. In January 1922, a fourteen-year-old boy, Leonard Thompson, was the first person with diabetes to receive insulin, saving his life and bringing his diabetes under control.

Insulin was first crystallised in 1926 but no one had yet studied its structure by X-ray crystallography. Initially, in the early 1930s, Dorothy found that the crystals were too small and painstakingly re-dissolved them and grew larger versions. From these, she was successful and her eureka moment arrived: the first X-ray photographs of the flower-like crystals of insulin. In 1935, Dorothy published her findings in *Nature*, her first sole-author paper at the age of twenty-five. However, it wasn't until some thirty-five years later, with the increasing use of techniques such as isomorphous replacement and more complex mathematical/computational analysis, that Dorothy was able to solve the full structure of the insulin molecule.

In the 1960s Dorothy had a large group of colleagues, with funding from both the Royal Society and the Science Research Council. Marjorie Aitkin, one of Dorothy's chemistry students, and Beryl Rimmer worked on the first good heavy atom derivative of insulin. They were using crystals provided by the Danish chemist Jørgen Schlichtkrull, who had an inherent interest in insulin as his daughter had diabetes. His crystals contained either two or four zinc atoms per six-molecule unit. This had the effect of confusing Dorothy and her team, as the images were very difficult to interpret, until they realised that the concentration of salt influenced which crystal was produced. If the chloride concentration was too high, the four-zinc version was produced.

Technical problems such as this were, and still are, common in crystallography. Interpreting the mass of data was also an issue until computers became more widely available and sophisticated. At every stage of her career, Dorothy made use of the latest computer. John Rollett had arrived in the laboratory in 1955 from the California Institute of Technology and became highly proficient at developing computer programs for crystallography analysis. In 1961 Eleanor Dodson (née Collier), who was a mathematician, also became a computer expert in Dorothy's laboratory.

Michael Rossman and David Blow, from Dorothy's friend Max Perutz's laboratory in Cambridge, had a significant role to play, too. Their research was the next milestone, after isomorphous replacement, in solving the structure of complex proteins. They had developed mathematical methods to study the close relationships between sub-units in crystals. Using methods of rotation and translation, they showed that insulin was a hexamer: the insulin molecule's six sub-units were arranged in pairs, related to each other by a two-fold axis (a rotation of 180 degrees about a line), at right-angles to the three-fold axis (anti-clockwise rotation of 120 degrees about a line).

It was a complicated structure and a seminal moment for those present. Dorothy, had solved the structure of penicillin with 39 atoms, vitamin B12 with 181 atoms, and now her largest molecule – insulin with 777 atoms. David Blow commented afterwards that he was very impressed by Dorothy's ability to visualise 3D structures in her mind from 2D data. Computers did play an increasingly vital role but, just like with vitamin B12, the work also advanced through Dorothy's intuitive understanding of the data in front of her.

Marjorie Aitkin's work was eventually published in 1966. Over the next few years, several factors allowed the laboratory to derive a more complete structure. Working with Mamannamana Vijayan from Bangalore, Guy Dodson, a biochemist married to Eleanor, obtained a total of 60,000 diffraction patterns from five different heavy chain derivatives. These were made using a new method, where he removed the zinc atom out of the insulin crystals and replaced the zinc with other metals. These lead and cadmium derivatives were then analysed in a new diffractometer, a very accurate and quick machine that had arrived in 1968.

The insulin structure was published in *Nature* in September 1969, in a paper with ten authors and acknowledging the contributions of twenty-three other scientists. This demonstrates both the long and circuitous path of discovery, involving a large number of scientists, and also Dorothy's desire to see everyone's contribution recognised. Unable to see other scientists as competitors, Dorothy listed anyone involved; the important thing was not who solved the problem but that the problem was solved. Today's scientific papers often list in excess of twenty authors, particularly in the field of genomics. In the 1960s, this was much less common. A number of these authors, including Guy Dodson, continued the studies of insulin for many more years.

As Dorothy became internationally recognised, she realised that she was in a unique position to influence and persuade. She had been inspired by her mother, and the lasting impression of their attendance, when Dorothy was a teenager, at the League of Nations General Assembly. Like Molly, Dorothy had friends worldwide. Dorothy's recognition of the power of communication and collaboration in science, combined with her non-prejudicial approach, meant that she was well placed to foster international relations and scientific interaction. As she explained, 'When the war ended, we decided the first thing that crystallographers had to have was an international union of crystallographers to encourage everyone to meet and exchange information.' It seemed a perfectly harmless aim but including communist countries proved very difficult.

In Cambridge, Dorothy had been influenced by the anti-war left-wing feeling that was prevalent in the 1930s. Her brand of socialism was not fervently ideological. Rather, it was personal, concerned with equal opportunities for all in their chosen field, science or otherwise, men or women. As a great traveller, she witnessed first-hand life in countries like Vietnam, Russia and China where she was drawn to the modest, hard-working communities, with good schools and hospitals. Dorothy was not immune to the tyranny of some communist leaders, although there is still some debate as to how much she, or others, knew of the worst excesses of such regimes at the time. But she was not afraid of being in the public eye and putting her viewpoint forward.

China was a regular destination for Dorothy. In the late 1950s, Chinese scientific research was almost non-existent. After Mao Zedong came to power in 1949, there was a drive to increase agricultural and industrial productivity, beginning with the first Five Year Plan in 1953 and then the Great Leap Forward in 1958. Science was increasingly sidelined as bourgeois and intellectual, unless it was seen to increase productivity. One exception to this was an insulin synthesis project, a token gesture on the behalf of the Chinese government to promote basic research.

On one of her eight visits to China, Dorothy was very taken by the enthusiasm of the scientists involved. Her personal visits were important, as the Chinese were not allowed, at that time, to publish their results in Western journals. In 1967, the Chinese solved the structure of insulin independently, two years after Dorothy. The achievement was not recognised by the Chinese government but Dorothy brought the result to the world's attention. Her interest in keeping the lines of communication open meant that, when Mao's regime fell after his death in 1976, the Chinese crystallographers were in a stronger position to participate internationally.

During the 1970s, as well as becoming president of the International Union of Crystallography and the Council of the British Association for the Advancement of Science, Dorothy headed up the Pugwash Conference. The latter, in many ways, epitomises the nature of Dorothy's work in both the laboratory and on the world stage. Like Lise Meitner (see Chapter Eight), Dorothy was passionate about scientific responsibility – particularly the need for scientists to acknowledge potentially dangerous or ethically challenging discoveries. The Pugwash Conference draws its inspiration from the manifesto, put together by Bertrand Russell and Albert Einstein in 1955, which urged world leaders to promote nuclear disarmament and to use peaceful methods to settle disputes. Its mission is to use scientific insight and technological development to solve problems, particularly in areas where science and world affairs meet.

Dorothy was not present at Pugwash's first meeting in 1957 in Canada but her close friend Kathleen Lonsdale encouraged her to attend the 1962 meeting in London. Over the following decade, Dorothy proved to be a great asset to Pugwash and in 1975, aged sixty-five, she became its president. It was clear that the organisation recognised that Dorothy would be more than a figurehead. With her strong networking skills, she was well known and trusted by her scientific friends in countries like Russia and China, and she had a passionate interest in the developing world and the ability to present ideas coherently and persuasively.

Behind the scenes, Dorothy was working towards the thawing of East–West relations, with her quiet insistence on keeping the lines of communication open. With the huge increase in support for the Campaign for Nuclear Disarmament in Britain, and similar groups elsewhere in Europe, the tide was turning. In 1983, Dorothy made use of an important contact: her former chemistry student Margaret Thatcher. She arranged to meet the Prime Minister at her country residence Chequers to discuss the relationship with the Soviet Union. Their politics may have been quite different but they respected each other's opinion and Dorothy's contribution was eventually recognised, not by the British, but by the Soviets who awarded her the Lenin Peace Prize in 1987, at the age of seventy-seven.

Dorothy was above all concerned about people. Whether promoting science or humanitarian ideals on the world stage, or working in the laboratory, creating the right working conditions was important to her. Many aspects of Dorothy's laboratory would be familiar to today's scientists; some aspects were distinctly different. Sivaraj Ramaseshan, who worked with Dorothy in the 1960s on insulin, was struck by the contrast between the formal nature of Indian laboratories and the informal, although highly productive, set-up of Dorothy's. He said, 'Whenever it was sunny we used to watch cricket, then go to a pub . . . From the point of view of relaxation and doing science it was one of the best periods of my life.' And one of Siv Ramaseshan's colleagues in Bangalore, M. A. Viswamitra, was equally complimentary about his experience in Dorothy's laboratory, saying, 'The family visits to her home were another thing that contributed to our success. Not that we were talking about crystallography – but somehow you would come back and feel that you could do even better.'

Everyone was called by their first names instead of the more common practice, at the time, of using surnames for men and Miss or Mrs for women. Although Dorothy did not see herself as a feminist, gender equality was a priority. On one occasion, Dorothy took up the cause of just pay for women after discovering that her student was to receive a cut to her annual grant on getting married. She wrote to the Department of Scientific and Industrial Research, 'I could never have carried out the amount of scientific research I have achieved if I had not, at the time of my marriage, been earning a sufficient salary to permit me to pay for help in our home.' Dorothy's appeal was successful and the original sum was maintained, although this proved to be an exception, rather than an official change in policy.

The welfare of university students and workers was always a principal concern of Dorothy's. In 1965, she founded Linacre College in Oxford for overseas graduate students. She was also instrumental in changing the archaic employment regulations for women. At the end of the 1960s, Dorothy was on a committee at the University of Birmingham, set up to investigate its administration. A number of women working part-time there, and in Oxford, had no proper contracts. One of them, Eleanor Dodson, said, 'She came back to Oxford and immediately wrote me a contract . . . she had been very underpinned by the fact that she had a proper job at Somerville . . . and that they paid her while she was having children.'

After Dorothy retired as a Wolfson professor in 1977, aged sixty-seven, she maintained a base in the crystallography department at Oxford where she read, wrote and advised passing students. Her enthusiasm and participation in various organisations continued with equal fervour. British universities were undergoing a period of change in the 1960s and 1970s, with the student voice increasingly evident. Bristol University was no exception. After a student 'thought Dorothy Hodgkin would be a good idea', she became the first non-royal female chancellor of a British university. She held the post from 1970 to 1988 and was a strong advocate for the students' causes. Among her many legacies in Bristol are Hodgkin House, a hostel for overseas students, and the Union's Hodgkin Scholarship for southern Africans.

Her legacy to science is what Dorothy will be most remembered for. She determined the molecular structure of penicillin in 1945, vitamin B12 in 1956, insulin in 1969 and countless other proteins through X-ray analysis. With each new discovery, Dorothy increased the size and complexity of the molecules she studied and advanced the field of crystallography.

The 1950s, in particular, was a heady time in the analysis of protein structure and its relationship with protein function. In 1954, Vincent du Vigneaud at Cornell University synthesised the hormone oxytocin – the first naturally occurring protein to be artificially made. In 1956, the 3D structure of proteins was linked to the sequence of its amino acids, so that by 1957 John Kendrew solved the first 3D structure of a protein, myoglobin – the oxygen-carrying protein found in muscle. This was followed in 1959 with Max Perutz's 3D structure of haemoglobin, twenty-three years after he had first started on the project.

Kendrew and Perutz were jointly awarded the Nobel Prize in 1962, two years before Dorothy. She was awarded the Nobel Prize not just for determining the structures of several vitally important molecules but for extending the bounds of chemistry itself. By choosing projects others considered impossible, Dorothy helped to establish one of the characteristic features of contemporary science: the use of molecular structure to explain biological function. And knowing a protein's structure allowed targeted research for drug development.

Dorothy's findings allowed others to grow and expand their research on protein structure at a time when X-ray crystallography, imported from the field of physics, was mistrusted as a valid technique, compared with more conventional chemical analysis. When penicillin's core was revealed as a ring of three carbon atoms and nitrogen, believed to be too unstable to exist, the scientific community reacted with incredulity. One chemist, John Cornforth, declared, 'If that's the formula of penicillin, I'll give up chemistry and grow mushrooms.' But her formula was correct and was the starting point for the synthesis of chemically modified penicillin. Dorothy's work on the penicillin molecule finished too late for the wartime synthesis of penicillin; yet the structural knowledge gained in the war years proved invaluable in developing penicillin-like antibiotics after the war that could be administered more conveniently, were more effective and had fewer side effects.

Vitamin B12 was the first organometallic compound to be identified and, like penicillin, its structure also revealed previously unseen features: the corrin nucleus consists of a strange ring of nitrogen and carbon atoms surrounding its central cobalt atom and a novel bond between these atoms provided the clue to the vitamin's biological function.

Nowadays, much of the work in crystallography is automated and takes a matter of hours or days. Then, the process took years and sometimes decades. Dorothy's trademark was a combination of intellectual vigour and intuition. Despite her achievements, she was modest and realistic, once saying to a journalist, 'You ought to realise that for 90 per cent of my life, I'm dealing with failure, and occasionally I have a success.'

Unlike other crystallographers of her generation, Dorothy was not associated with any particular technical breakthrough but she was always keen to make use of new computer technology and this eventually led to the structure of insulin. Now insulin's structure is known, biologists are focused on how it is made, which receptors it binds to and how it is transported around the body. Armed with this knowledge, genetic engineers can change the chemistry of insulin to improve its benefits for diabetics.

Dorothy was not only passionate about her own science but cared deeply about her students and their careers. Her team of international and UK scientists were like an extended family and Dorothy maintained close contact with them long after they had left. One described her as 'a teacher, mother, friend and guide all rolled into one'. Many of them were women: Jenny Glusker, Judith Howard, Pauline Harrison and Eleanor Dodson.

A number of Dorothy's protégées set up their own highly successful research laboratories. A key figure, Guy Dodson, became a lecturer and subsequently a professor in the Chemistry Department at the University of York in 1976. Like his mentor, Dodson was a past master at making contacts and collaborated closely with the pharmaceutical industry to prepare suitable variants of insulin to treat diabetes.

A year after Dodson died, in a 2013 *Nature* paper dedicated to him, scientists in his laboratory played a pivotal role in new research that signalled a significant step forward in the understanding of how insulin works. The research provided a first glimpse of the hormone receptor complex, showing that insulin undergoes a conformational change as it engages with its receptor and that key elements of the receptor also remodel. Ninety years after the discovery of insulin and forty-three years after Dorothy Hodgkin had determined its structure, it was a timely reminder of the power of X-ray crystallography and structural biology to visualise fundamental biological processes.

X-ray crystallography is more important than ever. By 2003, the human genome project had revealed the sequence of the three billion letters, or base pairs, which make up the complete set of DNA in the human body. The genes contained within this sequence are translated into proteins, the structures of which are still largely unknown. Proteins remain difficult to crystallise but new, automated and high-speed analytical equipment is available, like the Diamond Synchrotron (Didcot UK), which produces very intense, narrow X-ray beams. By understanding the structure of the complex molecules involved in, for example, Alzheimer's disease, motor neurone disease and cancer, suitable therapies for these intractable diseases are on the horizon.

As the significance of Dorothy's findings became clear, she and others were aware that a Nobel Prize was on the cards. There is often a prolonged time lapse, however, between the execution of a scientist's work and the award of a Nobel Prize. In 1964, after she had published the structure of vitamin B12 and penicillin, some eight and fifteen years earlier respectively, Dorothy Hodgkin was the sole recipient of the Nobel Prize in Chemistry.

Dorothy remains, in 2018, the only British woman to win a Nobel Prize in science. Only just over 2 per cent of chemistry Nobel Prizes have been awarded to women worldwide. The award was given 'for her determinations by X-ray techniques of the structures of important biochemical substances'. The British press focused on what they viewed as the most significant aspect of her award: her role as a housewife and mother. The *Observer*'s reaction was typical: the 'affable-looking housewife' had won the prize for a 'thoroughly unhousewifely skill: the structure of crystals of great chemical interest'. Dorothy saw no reason why she shouldn't combine all these aspects of her life successfully. Her ability to do so at the time was highly unusual.

A year later, in 1965, Dorothy was also awarded the Order of Merit. Only twenty-four people at any one time can hold this personal gift from the Queen, in recognition of a substantial contribution to the arts, sciences or public life; it is the highest honour any British citizen can receive. Dorothy's contribution to the world of crystallography lives on principally through her scientific achievements but also in several artistic ways. A number of portraits and sculptures exist, including the one by Maggi Hambling, painted in 1985 and housed in the National Portrait Gallery in London. It depicts a seventy-five-year-old Dorothy with multiple hands, cruelly gnarled by rheumatoid arthritis, still frantically working, too busy to finish a half-eaten sandwich, in her last home – Crab Hill in Warwickshire. The portrait also celebrates her scientific success, with a large ball-and-stick model of a molecule, surrounded by piles of papers.

Dorothy has also twice been featured on British stamps. In 1996, she was one of five women on a 'Portraits of Genius' series, featuring '20th Century Women of Achievement'. In 2010, the Royal Society (in collaboration with the Royal Mail) celebrated its 350-year anniversary by issuing a set of ten stamps. Chosen from more than 1400 fellows and more than 60 Nobel laureates, Dorothy was the only woman. In 2014, Google, as part of their series to highlight the achievements of women, celebrated Dorothy's birthday with a Google Doodle, where the Google 'Os' were the carbon rings of penicillin.

As Dorothy advanced in years, her large circle of friends and colleagues slowly diminished. Her friend Kathleen Londsdale, the physicist, died in 1971, followed shortly afterwards by the father of British crystallography, Sir Lawrence Bragg. The most devastating personal blow for Dorothy was the death of her beloved Thomas in 1982, aged seventy-two. His health, after years of heavy smoking, had been slowly getting worse. On their way home from a long trip to the Sudan, they stopped over in Greece. There, Thomas collapsed with heart failure and, three days later, he died.

Grief-stricken, Dorothy turned to her three children and seven grandchildren, and in particular her daughter Elizabeth. Dorothy listed 'children' as a recreation in her *Who's Who* entry. Jenny Glusker, one of her students, said, 'She gave the children a lot of attention, providing time to talk to them rather than fussing around the house. I learnt a lot from her about what was important and what you let slide.' She was very proud of her family and their success in their chosen fields, no doubt influenced by their talented and hard-working parents.

Luke studied mathematics at Balliol College and St John's College, Oxford. Now retired from a university career teaching mathematics and history, he works as a freelance writer and teacher. Recent publications include *A History of Mathematics: From Mesopotamia to Modernity*. After gaining a PhD in history, and as an Arabic speaker, Elizabeth Hodgkin taught medieval history at the University of Khartoum in the 1970s and maintains close links with the Sudan, where her mother lived as a child. She worked as a human rights researcher with Amnesty International from the late 1980s. Toby Hodgkin is a scientist in Italy, working in the area of agrobiodiversity, the agricultural biodiversity that is the result of natural selection.

Despite her increasingly debilitating rheumatoid arthritis, Dorothy still travelled, insisting on one last trip to China, one year before she died. She was now mostly confined to a wheelchair and her Chinese friends were shocked at how frail she looked. Returning home from this trip, she broke her hip for the second time. Dorothy's indomitable spirit was finally diminishing and she died at home surrounded by family on 29 July 1994, aged eighty-four.

Scientific biographers do not always find much correlation between good character and great science but Dorothy is one of the exceptions. In his obituary of her, Max Perutz wrote, 'Dorothy

Hodgkin's uncanny knack of solving difficult structures came from a combination of manual skill, mathematical ability and profound knowledge of crystallography and chemistry. It often led her, and her alone, to recognise what the initially blurred maps emerging from X-ray analysis were trying to tell. She will be remembered as a great chemist, a saintly, gentle and tolerant lover of people and a devoted protagonist of peace.'

CHAPTER SIX

Henrietta Leavitt
(1868–1921)

Henrietta Leavitt was an American with an influence that made her a star among astronomers. She discovered a way of ranking stars' magnitudes using photographic plates, which became a standard in the field. Henrietta also developed a method by which astronomers can accurately measure extra-galactic distances known as the period-luminosity relation. This enabled her to determine the distance to stars that are so far enough away that we can learn about the scale of our universe from them. So important was this work that, in 1923, it enabled Edwin Hubble to show that the great Andromeda Nebula, Messier 31, was much too far away to be part of our galaxy.

Henrietta was nominated for the Nobel Prize in 1926, but sadly she died five years earlier at the age of just fifty-three, so could not be considered for it as the Nobel Prize is never awarded posthumously. Although she received very little recognition during her lifetime, a fellow astronomer subsequently described her as a 'star fiend'. Since her death, her contribution to twentieth-century astronomy has been more widely recognised: the crater Leavitt on the Moon is named in her honour, as is the asteroid 5383 Leavitt.

Henrietta Swan Leavitt was born in 1868 in Lancaster, Massachusetts. Her father George Roswell Leavitt was a congregational minister and her mother was Henrietta Swan (née Kendrick). By the time of the 1880 census, the family was living at 9 Warland Street in Cambridge, Massachusetts, occupying half of a large double house. George was working as the pastor at the Pilgrim Congregational Church on the corner of Magazine and Cottage streets, just a few blocks away. By 1880, Henrietta's family had grown to include a younger brother, George, and three younger sisters. Martha, Caroline and Mira. Mira, as was so common in those days, would not survive to see her third birthday, and was reported in the 1880 census as having died. Another brother, Roswell, had died in 1873 when he was only fifteen months old. Henrietta's youngest brother, Darwin, would be born in 1882.

In the other half of the house, Henrietta's grandfather, Erasmus Darwin Leavitt, lived with his wife and his thirty-year old daughter. Henrietta's father had graduated from Williams College, a prestigious liberal arts college in the Boston area, and had earned a doctorate in divinity from the Andover Theological Seminary. In the early

1800s, the family moved to Cleveland Ohio, and in 1885 Henrietta enrolled at Oberlin College, where she took a preparatory course followed by two years of undergraduate study. After these two years, Henrietta moved back to Cambridge, where she enrolled at Radcliffe College in 1888. Radcliffe College was the undergraduate women's college affiliated with Harvard (which only admitted men at this time) and was considered one of the best women's colleges in the country.

Radcliffe's entrance requirements were among the strictest. Whatever subject a student aimed to concentrate on at Radcliffe, she was expected to be familiar with a list of the classics, including Shakespeare's *Julius Caesar* and *As You Like It*, Samuel Johnson's *Lives of the Poets*, Jonathan Swift's *Gulliver's Travels*, and Jane Austen's *Pride and Prejudice*. Applicants for entry were also expected to write, on the spot, a short composition, show proficiency in Latin, Greek, German and French, be knowledgeable about history (they had the choice of either the history of Greece and Rome or the history of the USA and England), and pass exams in mathematics (which included algebra up to quadratic equations and plane geometry), physics and astronomy. Quite a daunting prospect. In addition, students were required to show more advanced knowledge in two subjects, which they could choose. The Radcliffe College catalogue stated, 'A candidate may be admitted in spite of deficiencies in some of these studies; but such deficiencies must be made up during her course.'

The only deficiency shown in Henrietta's entrance exam was in history, and by her junior year she had demonstrated that she had corrected this. Most of her time at Radcliffe was taken up doing courses in Latin, Greek, the humanities, English, modern European languages (German, French and Italian), fine arts and philosophy. She was not offered much science; just natural history, an introductory physics class (in which she got a B), and a course in analytical geometry and differential calculus (she received an A). It was only in her senior year that she enrolled in an astronomy course, earning an A-. The astronomy classes were held at the Harvard College Observatory, just up Garden Street from Radcliffe College, and they were taught by the astronomers who worked there, supervised by the observatory's new director, Edward Pickering.

In 1892, just before her twenty-fourth birthday, Henrietta graduated from Radcliffe with a certificate that stated that she had completed a curriculum equivalent to what, had she been a man, would have earned her a bachelor of arts degree from Harvard. No BA degree could be awarded as she was a woman.

Edward Pickering had been appointed as director of Harvard College Observatory in 1877 when he was just thirty-one years old. The young and dynamic director was keen to put his stamp on the place. In 1885, after settling into his new position for a few years, Professor Pickering had the ambitious idea of cataloguing the position, brightness and spectrum of all the stars visible in the sky, down to ninth magnitude from both his observatory in Cambridge and from a southern observing station that Harvard had established in Arequipa, Chile.

The magnitude system used by astronomers for measuring the brightness of stars has its origins in the work of the ancient Greek astronomer Hipparchus, who lived from about 190 BC to 120 BC. From the island of Rhodes, Hipparchus classified stars based on how bright they appeared to his eyes, stating that the brightest stars were of the first magnitude, and the faintest he could see were of the sixth magnitude. It was a fairly arbitrary system until English astronomer Norman Pogson proposed a more mathematical definition in 1856. He stated that a star of the first magnitude would be exactly one hundred times brighter than a star of the sixth magnitude.

The magnitude system is therefore a negative system. A star of the first magnitude is brighter than a star of the sixth magnitude. Ninth magnitude stars are fainter than sixth magnitude stars, and stars such as Sirius that are brighter than a first magnitude star have negative magnitudes. The system is defined by the star Vega, which has a magnitude of 0.

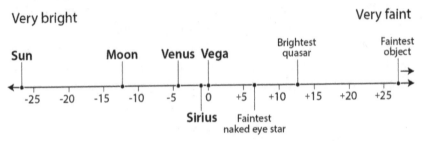

Apparent brightness of some objects in the magnitude system

The other confusing thing about the magnitude system is that it is logarithmic. This means that a magnitude difference of 10 does not mean a difference in brightness of 100 + 100, but rather of 100 x 100 = 10,000. A magnitude different of 15 would be 100 x 100 x 100 = 1,000,000.

The faintest stars one can see with the naked eye, in a dark enough place, have a brightness of about sixth magnitude. Sirius, the brightest star in the night sky, has a magnitude of -1.46. Ninth magnitude stars are nearly sixteen times fainter than sixth magnitude stars, far too faint to be seen with the eye, and so would only be visible on the photographic plates being taken at Harvard and in Chile.

Much of the funding for the huge undertaking that Professor Pickering wished to embark upon came from the widow of Henry Draper, who had been a pioneer of astrophotography. Henry Draper was a physician by training, but his real passion was astronomy. In 1872, he photographed the first ever spectrum of a star that was not the Sun when he obtained the spectrum of Vega; it showed the numerous absorption lines that Joseph von Fraunhofer had first noticed in the Sun's spectrum in 1814–15. Although Henry Draper was dean and professor of medicine at New York University, in 1873 he resigned his position to pursue his passion for astronomy on a full-time basis. Having married wealthy socialite Mary Anna Palmer in 1867, he was in the fortunate position of not having to worry about money.

Henry Draper was also the first person to photograph the famous Orion nebula, the object that forms the central 'star' of Orion's sword. He died in 1882 at the age of just forty-five from a double bout of pleurisy, but by the time of his death he had obtained the spectra of over one hundred stars. In his honour, his widow decided to donate a large sum of money to Harvard, which funded a telescope and much of the work of Professor Pickering's ambitious stellar catalogue. The huge catalogue became known as the 'Henry Draper Catalogue'; it was not completed until 1924, shortly after the professor's own retirement. It gave the position, brightness and stellar spectral classification for 225,300 stars.

As part of the enormous work necessary to catalogue so many stars, Professor Pickering employed a number of women who became known as the 'Harvard Computers', or 'Pickering's Harem'. Until the advent of electronic computers, the word 'computer' was used for anyone whose job entailed a lot of calculation, and it was quite common for scientific research establishments to employ women as computers as they were cheaper to employ than men.

The first woman Professor Pickering employed at the Harvard College Observatory as a computer was his housemaid, Williamina Fleming. She had immigrated to the United States from Scotland with her husband in 1878, but after giving birth she was abandoned by him and ended up working for the professor. He recognised her ability, and in 1881 he decided to offer her a position at the observatory, teaching her how to analyse stellar spectra. Williamina was quickly joined by several other women, including Annie Jump Cannon in 1896, Antonia Maury in the mid-1890s and Henrietta Leavitt in 1893.

Henrietta was assigned the task of determining the magnitude of many of the thousands of stars on the photographic plates that were being shipped from Chile and also being taken on the telescope just above the office that she shared with the other Harvard Computers at the observatory. Such was the huge amount of work involved in compiling this stellar catalogue that even the work of determining stars' magnitudes was split among several of the women. Henrietta's particular assignment was to measure the brightness of variable stars, stars whose brightness is not constant, but rather gets brighter and dimmer.

In the plates coming back from Chile, she concentrated on looking for variable stars in the Magellanic Clouds, the two large patches of light that were first noted by Portuguese and Dutch astronomers when they sailed south around the southern tip of Africa in the fifteenth century. They were initially known as 'Cape Clouds', and are only visible from places south of the Equator as they lie so far south in the sky. They were also noted in 1503–4 by Amerigo Vespucci, the Italian explorer after whom the Americas are named; and were described in detail by Antonio Pigafetta during Ferdinand Magellan's circumnavigation of the globe in 1519–22. Ever since Pigafetta's detailed descriptions, they have been known as the Magellanic Clouds.

The Large Magellanic Cloud (LMC) has a diameter of about 10 degrees, which is about twenty times the diameter of the full Moon. The Small Magellanic Cloud (SMC) is smaller on the sky, with a diameter of about 5 degrees (ten times the diameter of the Moon). From a dark place south of about the Tropic of Capricorn, they are high in the sky and quite easily visible. We now know that they are small satellite galaxies of our own Milky Way galaxy, but this was not known at the time that Henrietta was assigned the task of measuring the magnitudes of the variable stars in the two clouds.

As Jeremy Bernstein, the American physicist, later wrote, 'Variable stars had been of interest for years, but when she [Henrietta] was studying those plates, I doubt Pickering thought that she would make a significant discovery, one that would eventually change astronomy.' Henrietta noted thousands of variable stars in the SMC. She spent day after day painstakingly poring over the photographic plates. One colleague later referred to her dedication to her work as 'an almost religious zeal'. After two years of working on these variable stars, she wrote up her findings and then, in 1896, she sailed to Europe and spent two years travelling there.

When she finally got back to Boston she got in touch with Edward Pickering, who suggested some revisions to the work she had done before departing on her European trip. She took the manuscript she had written with her, and left for Beloit, Wisconsin, where her father had become the minister in another church. She remained in Beloit for two more years due to 'personal problems' that she never fully explained. During this time she worked as an art assistant at Beloit College, but clearly the work there did not satisfy her. She had no contact with Edward Pickering for a few years.

On 13 May 1902, she wrote to Professor Pickering, apologising for not getting on with the research that he had assigned her, and for not being in touch in such a long time. She said that she hoped she could continue with her work for him from Wisconsin. As she explained in her letter, 'The winter after my return was occupied with unexpected cares. When, at last, I had leisure to take up the work, my eyes troubled me so seriously as to prevent my using them so closely.' She assured Professor Pickering that her eyes were now strong, apologised for her absence and said that she was ready to resume work on determining the magnitudes of variable stars. Henrietta said that her enthusiasm for astronomy had not diminished, and asked if he could send her the notebooks that she would need to complete the manuscript.

She must have been relieved and delighted by the astronomer's response. Three days later he offered her a job. He told her that he would be willing to pay her 30 cents an hour 'in view of the quality of your work, although our usual price, in such cases, is 25 cents an hour'. If it were not possible for her to move back to Cambridge, he would pay her fare for a short visit to the Harvard College Observatory, and she could take what she needed back with her to Wisconsin. Henrietta accepted Edward Pickering's offer, and told him that she planned to arrive to take up her job before the first of July 1902. After a delay due to the illness of a relative in Ohio, Henrietta arrived in the Boston area on 25 August 1902 and worked through the autumn before taking another lengthy trip to Europe. Upon her return, she decided to relocate to Cambridge as a permanent member of staff.

By 1904, Henrietta's dedication to her measurements of the brightness of variable stars paid dividends. One spring day, she was comparing photographic plates of the SMC which had been taken at different times. She noticed several variable stars, and as she examined other images she found dozens more. During the autumn of 1904, sixteen more photographic plates of the SMC were taken at the observing station in Arequipa and shipped to the observatory in Boston. They arrived in January 1905, and Henrietta began studying them immediately. She found more and more variable stars, 'an extraordinary number,' she would later say.

Henrietta wrote up the results in April 1905's *Harvard College Observatory Circular* in a paper entitled '843 New Variable Stars in the Small Magellanic Cloud'. Unusually for that time, she was the sole author on the paper. It was the custom for the director of an observatory or research group to author papers, with the person who had actually done the work credited as a co-author or merely mentioned in the paper. The fact that Henrietta was sole author on this paper shows the high regard in which Edward Pickering held her and her fellow Harvard Computers. A Princeton astronomer wrote to Professor Pickering, remarking, 'What a variable-star "fiend" Miss Leavitt is . . . One can't keep up with the roll of the new discoveries.'

By this time Henrietta was boarding with her uncle Erasmus in a large Italian-style villa (now part of the Longy School of Music) on Garden Street, just a few hundred metres from the observatory. Being so close to work allowed Henrietta to spend long hours at the observatory, where she continued to catalogue variable stars in the two Magellanic Clouds. In 1908, four years' worth of Henrietta's work were written up by her in a paper entitled '1777 variables in the Magellanic Clouds', published in the *Annals of Harvard College Observatory*.

On page 107 of this paper Henrietta noted, based on data for sixteen stars, 'It is worthy of notice that in Table VI [giving periods of variables in the SMC] the brighter variables have the longer periods.' Given what we now know, this simple line is an understatement of huge proportions, similar to the one made by James Watson and Francis Crick at the end of their 1953 paper on the structure of DNA: 'It has not escaped our notice that the specific pairing we have postulated immediately suggests a possible copying method for genetic material', a coded sentence that meant they had discovered the secret of life.

Henrietta was not being modest; she just didn't want to over-interpret her data. She knew that more work was needed, and that is certainly something that Professor Pickering would have pointed out to her: sixteen stars were just not enough upon which to base any bold conclusions. Unfortunately, the 'further work' was interrupted; later in 1908, Henrietta fell ill. On 20 December, she wrote to Edward Pickering from a Boston hospital, where she had been for the previous week, thanking him for the 'beautiful pink roses' and 'for the kind thought so beautifully expressed. It means much at a time like this to be made to realize that one is remembered by one's friends.'

In order to recover, she returned to Wisconsin to be with her family. By now, her brother George was a missionary and her youngest brother Darwin was a clergyman like his father. She rested in

Beloit for the spring and summer of 1909, and hoped to resume her work in the autumn. In September, however, she wrote to Edward Pickering that a 'slight illness', which she had contracted after a visit to a lake near her home, had 'proved unexpectedly obstinate, and I cannot tell when I shall be able to get away'.

In October 1909, by which time she had been absent from work for nearly a year, Edward Pickering wrote to her to enquire whether she would like him to send her some work. By December she still had not responded, so he wrote to her again, and this time he started to show a little impatience with her, saying, 'My dear Miss Leavitt, it is with much regret that I hear of your continued illness. I hope you will not undertake work here until you can safely do so. It may however relieve your mind if we can dispose of two or three questions . . .'

His first question was a request for her to send a letter to him at the beginning of each month letting him know whether she would be returning anytime soon. His second request was whether she could give him a preliminary report on the results of another study he had asked her to do, the 'North Polar Sequence'. This was an attempt to measure, more accurately than ever before, the magnitudes of ninety-six stars near the North Star, Polaris. This was one of Edward Pickering's pet projects, in fact of higher priority to him than the work Henrietta had been doing on the variables in the Magellanic Clouds. He wanted to use the North Polar Sequence as the basis for gauging the brightness of all the stars throughout the sky.

Henrietta replied a few days later and apologised for being too weak to answer his letter of October, but she hoped that her condition would improve enough to resume work after Christmas 1909. In the event, she was not well enough to return to work until early in 1910, and she could only do so from Beloit rather than Cambridge. Henrietta was sent a box of photographic plates, paper prints, ledgers, a wooden viewing frame and a 1½ inch eyepiece and began sending detailed reports back to the observatory.

Henrietta was finally well enough to return to Cambridge on 14 May 1910 but her return was short-lived. In March 1911 her father died, and Henrietta returned to Beloit to be with her mother. In June, Professor Pickering sent her a box of seventy photographic plates and other material for her to continue work on the North Polar Sequence, and she took some of the material with her when she and her mother departed to stay with some relatives in Des Moines, Iowa. There, she also found time to continue her work on the variable stars in the Magellanic Clouds. Finally, she had some uninterrupted time to devote to the variable stars in the two clouds, and plotted twenty-five of them on a graph with the apparent brightness on the vertical axis and the period of variation on the horizontal axis. This plot showed her important discovery: the period-luminosity relationship for Cepheid variables (stars that have a regular cycle of brightness). The results were published in 1912 in the *Harvard College Observatory Circular*, but this time under Edward Pickering's name and not hers. The paper is entitled 'Periods of 25 Variable Stars in the Small Magellanic Cloud' and the opening sentence reads, 'The following statement regarding the periods of 25 variable stars in the Small Magellanic Cloud has been prepared by Miss Leavitt.'

Towards the bottom of the first page, the paper states: 'A remarkable relation between the brightness of these variables and the length of their periods will be noticed. In H.A. 60, No. 4 [the 1908 paper referred to on page 161], attention was called to the fact that the brighter variables have the longer periods, but at the time it was felt that the number was too small to warrant the drawing of general conclusions. The periods of 8 additional variables which have been determined since that time, however, conform to the same law.'

How did Henrietta know that the Cepheids that appeared to be brighter were, indeed, intrinsically brighter? When we look at the stars in the sky, the stars that appear brightest may not necessarily be the brightest; how far away they are will also affect how bright they appear. For example, Sirius appears to be the brightest star in the night-time sky; it has a magnitude of -1.41. But, it turns out, the blue star Rigel in the north-east corner of Orion (near enough to Sirius to be visible at the same time), which appears to have a magnitude of only 0.13 (about four times fainter), is, in fact, intrinsically much brighter. It only appears to be fainter because it is much further away, about 860 light-years distant compared to Sirius, which is only 8.6 light-years away, one hundred times closer. So, to be able to compare the intrinsic brightness of stars we need to measure how bright they appear to be, but we also need to know the stars' relative distances.

Had Henrietta been studying Cepheid variables scattered throughout the sky, she would not have been able to make her assertion that the intrinsically brighter stars took longer to vary their brightness as she would have had no idea how the different distances of the stars should be factored in. All the stars she was studying were in the LMC, however, so she was correctly able to assume that they were all at roughly the same distance. This was crucial. It meant that any Cepheid that appeared to be brighter than another one was *also intrinsically brighter*, and this was key to her discovery.

Just noting that Cepheid variables that took longer to vary their brightness were intrinsically brighter than those that took less time would not, in itself, have been a significant discovery. But, as we shall see, it allowed astronomers to measure distances that otherwise they had no way of doing. Measuring distances in the heavens has always been one of the biggest challenges to astronomers. How big or how bright something in the heavens appears is no indication of distance, as we have seen with the example of Sirius and Rigel. The Sun only appears to be brighter than Sirius and Rigel because it is much, much closer, only 8 light-minutes away. But, how do we actually know the distance to the Sun?

In fact, the distance from the Earth to the Sun was not known until the mid-1700s, and the person who came up with the method of doing it was the English astronomer Edmond Halley (he of the comet). In 1676, at the age of just nineteen, Halley was sent to the island of Saint Helena in the southern Atlantic Ocean by John Flamsteed, the first Astronomer Royal, in order to start cataloguing the stars of the southern skies. While there, in November 1677, he observed the planet Mercury moving across the disk of the Sun, something called a 'Transit'. Because they lie closer to the Sun than Earth, only Mercury and Venus can be seen transiting the Sun, but both are rare events. Transits of Mercury happen roughly thirteen to fourteen times a century, but transits of Venus are even rarer; they happen in pairs separated by eight years, but there is a gap of either 105.5 or 121.5 years between each pair of transits.

Halley was fully aware of the problem of astronomers not knowing the distance from the Earth to the Sun because it was possibly the biggest problem in astronomy in the 1600s. While watching Mercury move across the disk of the Sun in 1677, he realised that someone seeing the same event from, for example, London would see a slightly different view than the one he was seeing from Saint Helena in the southern Atlantic Ocean. This is the well-known phenomenon of parallax. If we look at the same thing from two different viewpoints, it will appear to shift against the background. Try holding a pencil in front of your face about 30 centimetres away and look at it through just your left eye. Now switch to looking at it through just your right eye; the pencil will appear to move against the background because you are seeing it from two slightly different viewpoints. This is how our two eyes give us a perception of depth in the world around us. With the transit of Mercury, Halley deduced that if a person elsewhere saw it taking a slightly different path across the disk of the Sun compared to what he was seeing, and if the difference between those two paths could be measured, then simple trigonometry would enable astronomers to work out the distance to the Sun.

When he got back to London and worked out the details, however, Halley realised that Mercury was too far away from Earth for the technique to work. The path taken by Mercury across the disk of the Sun as seen from two different locations on the Earth would just be too small to measure. Venus is a lot closer to Earth, so the technique would indeed work using a transit of Venus. Halley published this work in 1716, but he knew full well that the next transit of Venus was not until 1761, a date that he would not live to see.

Thankfully his idea lived on after his death, and European scientists made plans to use his method finally to measure the distance from the Earth to the Sun during the June 1761 and 1769 transits of Venus. With the results of these, in 1771, Thomas Hornsby, professor of astronomy at Oxford University, announced that the distance from the Earth to the Sun was 93,726,900 miles (150,838,824 kilometres), which is within 1 per cent of the currently accepted value.

Once the distance to the Sun was known, would it be possible to work out the distances to other stars? Surely, as we orbit the Sun, we should see nearby stars shift their position compared to more distant ones, again due to the effect of parallax. The trouble was that no one had ever seen such a shift in the position of any star. When Galileo proposed that the Earth was not the centre of the universe but, instead, went around the Sun, the fact that no one had seen any stars shift due to parallax was used as a strong argument against his idea.

As telescopes improved, astronomers continued to try to measure stellar parallax. Finally, in 1838, German astronomer and mathematician Friedrich Bessel managed to measure the stellar parallax of 61 Cygni, a faint star in the constellation Cygnus. He measured a shift in its position of 0.31 arc-seconds, which is 0.000086 of a degree. No wonder it had taken so long to see any stellar parallax: the angles were truly tiny. This is the angle 1 centimetre makes if it is placed about 20 miles away.

By the end of the nineteenth century, astronomers had been able to measure the stellar parallax for only several dozen of the closest stars. More distant stars were just too far away to exhibit any measurable parallax, so astronomers found themselves in a position similar to the 1600s; they did not have a way to determine the distances to the more distant stars, only the ones in our solar neighbourhood. Astronomers were stuck. For example, it meant that we did not know the scale of our Milky Way galaxy; estimates varied wildly from 10,000 light-years to 300,000 light-years across. Henrietta's breakthrough in 1912 was to provide a way to determine the distances to stars that were too far away to use stellar parallax; all that was needed to be able to use it was to determine the distance to a nearby Cepheid variable using the stellar parallax technique.

This was done the following year by Danish astronomer Ejnar Hertzsprung. In 1913 he determined the distances to several Cepheids using stellar parallax, and from measuring how long their luminosities took to vary he was able to calibrate Henrietta's period-luminosity relationship. With this calibration, the distance to any Cepheid variable could be determined, merely by measuring how long it took to vary its brightness and by measuring how bright it appeared to be compared to the brightness of the Cepheids for which Hertzsprung had determined distances. When this was done for the Small Magellanic Cloud, it was found to be nearly 200,000 light-years away. This was the first piece of evidence that it may not lie in our Milky Way, unless one assumed that our galaxy was towards the large end of the estimates of its size.

Probably the most fundamental shift in our understanding of the universe, after the realisation that the Earth was not the centre, was to come about as a direct result of Henrietta's work. For centuries, astronomers had noted numerous fuzzy patches in the sky, which were clearly not stars. The term used for such patches is nebula, which comes from the Greek word for 'cloud'. The Orion nebula is an example of what became known as a bright nebula. Astronomers had also noted planetary nebulae, which appeared in early eighteenth-century telescopes to resemble Saturn. We now know that they are nothing to do with planets; they are stars like our Sun coming towards the end of their lives and throwing off their outer layers.

Possibly the most puzzling type of nebulae seen were spiral nebulae. These looked like swirls of gas in the sky, and many astronomers assumed that they were indeed clouds of gas within our Milky Way, which were in the process of forming stars and a surrounding planetary system. Others argued that they were great systems of stars beyond our Milky Way, and the German philosopher Immanuel Kant coined the term 'island universes' to describe them. Throughout the nineteenth and early twentieth centuries, the debate raged as to whether these spiral nebulae were island universes or were star-forming regions within our Milky Way galaxy.

As telescopes grew bigger and photographic plates became more sensitive, it became possible to take long-exposure images of these spiral nebulae. One of the most detailed studies in the early twentieth century was done by probably the century's most famous astronomer, Edwin Hubble. He had studied spiral nebulae for his doctoral thesis at the University of Chicago's Yerkes Observatory. He used a 24-inch aperture telescope, which was ideal for photographing these objects, and in 1917 he submitted his thesis entitled 'Photographic Investigations of Faint Nebulae'. In 1919, after briefly serving as an army officer in the First World War, Edwin was given a position at Mount Wilson Observatory in southern California, which had been established in the early 1900s by George Ellery Hale, the Yerkes Observatory's first director.

George Hale had no equal in raising money for building large telescopes, and after completing the 60-inch reflector in 1908 he started raising money for what would become the 100-inch. The main source of money was from John D. Hooker, a Los Angeles-based industrialist, and when it went into operation in 1917 the Hooker 100-inch was, by far, the largest telescope in the world, gathering nearly three times more light than its next nearest rival. With its clear skies and incredible observing facilities, the best and the brightest astronomers were flocking to Mount Wilson Observatory, and Edwin Hubble was among them.

Edwin proved himself to be a hard-working and diligent observer, and was able to climb his way up the pecking order at Mount Wilson. By 1922, he was being given more and more time on the 100-inch, and he started studying many of the faint nebulae that he had studied in this thesis, using the larger telescope and better photographic plates. One of the nebulae that he studied was the great Andromeda Nebula, or Messier 31 as it is often known. Messier 31 is the only spiral nebula visible to the naked eye; if you go to a dark enough place in the northern hemisphere in the August–December period it is to be found in the constellation Andromeda just off to the north-east of the square of Pegasus.

Edwin was taking photographic plates of Messier 31 on a regular basis, and one night on one of his plates he noticed three stars that he had not seen before. He marked them with an 'N' for 'nova', meaning a new star. When he compared the plate with others that he had been taking, however, he saw that one of the stars was not new at all; he could see it on previous plates that he had taken. He could barely contain his excitement; this star was a *variable* star, a Cepheid variable. He crossed out the 'N' on the plate and wrote 'VAR!', realising that he could use it to measure the distance to the Andromeda Nebula using the period-luminosity relationship that Henrietta had found a decade before.

When Edwin did this, he could not believe what his calculations told him. He found the nebula to be at a vast distance, some two million light-years away. Surely he had made a mistake. He searched back through all the plates that he had of Messier 31 to see if he could find any other variables and he was in luck, he found another two. When he calculated the distances to these other two he got the same huge distance, two million light-years. Even with the high-end of estimates of the size of the Milky Way galaxy, such a massive distance put the Andromeda Nebula much too far away to be within our galaxy. It had to be an 'island universe'. Overnight, our understanding of the scale of the universe had changed.

We now knew that our galaxy was not the only one in the universe: Messier 31 and all the other spiral nebulae were galaxies in their own right, each containing hundreds of millions of stars. Later in the same decade, Edwin Hubble went on to show that more distant spiral galaxies were moving away from us more quickly; he had discovered that the universe is expanding. The method he used to determine the distances to these galaxies was, again, Henrietta's period-luminosity relationship for Cepheid variables.

By the time that Henrietta's work on the variable stars in the SMC was published, she was immersed in continuing work on the North Polar Sequence. For four more years she laboured away on this pet project of Edward Pickering's, except that there were frequent gaps for illness, sometimes for several months. In spring 1913, she was absent for three months but, by January 1914, she had finally finished this Herculean task of determining the brightness of the ninety-six stars in the North Polar Sequence. The work was published in 1917 in the *Annals of the Astronomical Observatory of Harvard College* in a paper that was 184 pages long. The paper was a masterpiece of diligent work, combining data from 299 photographic plates taken on thirteen different telescopes. Every stellar magnitude had to be checked carefully and cross-checked with its magnitude on other plates. It was work in which she could take great pride – PhDs have been awarded for much less.

Another astronomer, Harlow Shapley, was trying to determine the underlying reason for the variability of Cepheids. Working at Mount Wilson Observatory, he focused on gaining a fuller understanding of Cepheid variables, as he felt they could be used to determine the size of the Milky Way galaxy. In the mid-1910s, his work on Cepheids in globular clusters enabled him to show that the Sun was not at the centre of the Milky Way, and that the Milky Way was many thousands of light-years across. During this period, Harlow corresponded regularly with Henrietta, asking her for the latest data on Cepheids from her work on the Magellanic Clouds.

Henrietta's uncle Erasmus died in 1916, and she moved into a rooming house in Cambridge, where she lived alone. By 1919, Henrietta's mother had moved to Cambridge, and they were both living in an apartment building on the corner of Linnaean Street and Massachusetts Avenue, several blocks from the Harvard College Observatory. Although variable stars still occupied most of her working time, she was also looking for guidance on new projects. With Edward Pickering's death in February of that year, she was increasingly turning to Harlow Shapley for advice. In 1920 she wrote to him, asking him what she should work on next. Not surprisingly, he replied that it would be of 'enormous importance in the present discussion of the distances of globular clusters and the size of the galactic system' if she would work on plotting the periods of some of the dimmer variable stars in the SMC, those 'just fainter than the faintest already studied'. This is exactly what Harlow had been pestering Edward Pickering to get Henrietta to do in the months before his death.

Harlow also wanted her to see if the same period-luminosity law held for the variables in the Large Magellanic Cloud. In some ways, although he was giving Henrietta advice, Harlow was also treating her like a research colleague. In just a few months, however, he would become her boss. Following Professor Pickering's death, the astronomical community were on tenterhooks to see who would become the new director of Harvard College Observatory, one of the most sought-after positions in astronomy. After the favoured candidate turned it down, and disregarding concerns that Harlow Shapley was too brash and immature for the position, Harvard agreed to try the young astronomer on a one-year probation. In the spring of 1921, a few months after he had turned thirty-five, Harlow moved to Cambridge to take over where Edward Pickering had left off. By now, Henrietta was head of the stellar photometry group, which was determining the brightness of stars for the vast catalogue Edward Pickering had started many years before. The Henry

Draper Catalogue would ultimately fill nine volumes and give the position, magnitude and stellar classification for more than 225,000 stars.

During the decades that the Harvard women had been working at the Observatory, their status within Harvard had slowly improved. When Edward Pickering had tried to get Annie Jump Cannon an academic appointment, however, he had no success at all. The women were praised, somewhat condescendingly, for being 'good at detailed work'. Edward and others were on record as saying that their ability to do detailed work was because their minds were 'too simple to be distracted by thinking about other things'! Women were useful, but only allowed to do routine work, and Henrietta's contribution to higher-level astronomical research was unusual. Deeper matters, requiring thought and invention, were reserved for the men. Vassar College graduate Antonia Maury complained to a friend, 'I always wanted to learn the calculus, but Professor Pickering did not wish it.' Harlow Shapley was also a man who would measure how hard a computational task was in units of 'girl-hours', or if it was a really large task in 'kilo-girl-hours'.

Whatever Henrietta may have hoped for under Harlow's new directorship, her time under him was to be short-lived. By late 1921, she was sick again, but this time with cancer. Annie Jump Cannon wrote in her diary on 6 December, 'Went to see poor Henrietta Leavitt, dying with a malignant stomach trouble. So thin & changed. Very, very sad.' Less than a week later, on 12 December, Henrietta died, aged fifty-three. She was buried at Cambridge cemetery in the Leavitt family plot. A few days before her death she wrote out her will; the total value of her estate (which consisted of odds and ends such as a rug, a table and a bedstead) was $314.91.

Four months after Henrietta's funeral, Annie Jump Cannon was on a steamer bound for Peru. She was going on a tour of the Andes and to visit the remote observing station in Arequipa. One evening she made a note in her diary, 'Magellanic Cloud (Great) so bright. It always makes me think of poor Henrietta. How she loved the "Clouds".' Her colleague at the Harvard College Observatory, Solon Bailey, also remembered her fondly, writing in an obituary:

Miss Leavitt inherited, in a somewhat chastened form, the stern virtues of her puritan ancestors. She took life seriously. Her sense of duty, justice and loyalty was strong. For light amusements she appeared to care little. She was a devoted member of her intimate family circle, unselfishly considerate in her friendships, steadfastly loyal to her principles, and deeply conscientious and sincere in her attachment to her religion and church. She had the happy faculty of appreciating all that was worthy and loveable in others, and was possessed of a nature so full of sunshine that, to her, all of life became beautiful and full of meaning.

Her serious application to her science was celebrated by her peers. When the International Astronomical Union held its first general assembly meeting in Rome in May 1922, the Commission of Stellar Photometry, of which Henrietta had been a member, applauded her 'great service to astronomy . . . She was one of the pioneers in a difficult field of investigation in which she worked with conspicuous success, and it is deeply regretted that she was unable to finish her last undertaking [the North Polar Sequence].'

One of the most fitting testimonies to the value of Henrietta's period-luminosity law was the launch of the Hubble Space Telescope in 1990. With this powerful telescope, it was now possible to observe Cepheid variables as far away as in the galaxies in the Virgo cluster, too far away to be seen with ground-based telescopes. Now it was possible to measure the expansion rate of the universe.

The universe encompasses everything in existence, from the smallest atom to the largest galaxy. Since forming some 13.7 billion years ago in the Big Bang, the universe has been expanding and may be infinite in its scope. When scientists talk about the expanding universe, they mean that it has been growing ever since its origins. The part of the universe that we know about is called the observable universe, the region around Earth from which light has had time to reach us.

Imagine the universe is like a loaf of raisin bread dough. As the bread rises and expands, the raisins move farther away from each other, but they are still stuck in the dough. In the case of the universe, there may be raisins, or stars, out there that we can't see any more because they have moved away so fast that their light has never reached Earth. We know that our universe is expanding but how fast and what drives it is still a mystery, yet with Henrietta's period-luminosity law and the Hubble Space Telescope, new measurements were possible.

In 2001, the results of this work were announced; the universe was found to be expanding at a rate of 72 kilometres a second per megaparsec (roughly 3 million light years) – a measurement that would have been impossible without Henrietta's contributions to astronomy. Now new studies of supernovae (exploding stars) in remote galaxies and a force called dark energy are being conducted to understand the possible fates of the universe. Henrietta's name is rarely heard outside this rarefied world, yet her findings have immeasurably enhanced our understanding of the true scale of the universe and its origins.

Rita Levi-Montalcini
(1909–2012)

Rita Levi-Montalcini's wartime experiments in her hidden laboratory revealed the existence of nerve growth factor (NGF): the first growth factor to be discovered, which provided a vital insight into the growth and development of the nervous system. Emerging from the shadow of her domineering father and the challenging conditions of wartime Italy, Rita forged a career in medicine and the field of neurobiology. Before efficient protein sequencing and recombinant DNA techniques were available, the painstaking work to isolate, identify and characterise NGF took almost twenty-five years.

Rita's discovery and characterisation of NGF earned her a Nobel Prize, led the way to a multitude of other growth factors and revolutionised the field of embryology. Her work helped to explain how a complex organism can grow and develop from a single cell through a sequence of chemical communications, and has contributed to the understanding of pain control, Alzheimer's disease and cancer.

After a delayed start to her scientific career, Rita's unstoppable drive to succeed was matched by her longevity. Living until she was 103, she combined charm and an elegant, manicured appearance with a fierce desire to make a difference in the world.

Rita Levi-Montalcini was born on 22 April 1909 in Turin, northern Italy. This industrial city, not far from the border with Switzerland, is the capital of the Piedmont ('foot of the mountains') region and sits alongside the Po, Italy's largest river. It was a grand place to be born, with a cultural centre that rivalled many of Italy's other historic cities.

Rita's father, Adamo Levi, was an electrical engineer and a highly successfully factory owner, who provided a comfortable existence for his family. He was a strong personality with a commandeering presence and Rita was in awe of him. In her autobiography, *In Praise of Imperfection*, she recounts how she recoiled from his kisses, in fear of being scratched by his facial hair. His moustache matched his bristly personality and his outbursts of anger terrified Rita and her siblings. Even though she didn't always find favour with her father, she admired him. Rita was timid and retiring as a child but, as she grew up, the steely determination and strength of character she had inherited from her father gradually emerged. They were kindred spirits in many ways and, later in life, she commented on their shared 'tenacity, energy, ingenuity and dedication to work'.

The Levis were Jewish but not strict observers of Judaism; Adamo Levi encouraged his children to call themselves freethinkers. Rita's parents had many Catholic friends and found the Italian people of the secular state of Piedmont tolerant and welcoming. This liberal attitude was in sharp contrast to countries like Russia where Jews had been so violently treated in ghettos. Adamo told her about the pogroms in Odessa in 1905, and Rita often had nightmares about what might happen to her family.

Rita's mother Adele was also a profound influence and, as an adult, Rita distinguished herself from the other intellectual Levis of Turin by changing her name to Levi-Montalcini, encompassing her mother's maiden name. Adele was nine years younger than her husband and had been brought up to be a traditional Italian wife and mother. Adamo ruled the roost at home, so much so that the gentle Adele never questioned his decisions. This was not to Rita's liking, who adored her mother, and Rita feared that she too was destined to be tied to the hearth. Despite her misgivings about the Victorian way the household operated, Rita described her childhood as 'filled with love and reciprocal devotion'.

There were four children in the family. The eldest, Anna and Gino, were five and seven years older than Rita respectively. This age gap meant Rita was less close to them but she admired and looked up to them. Gino trained to be an architect, which he excelled at, becoming one of the most prominent Italian architects of the post-war period. Anna had dreamt of being a writer but found herself fully occupied in the traditional role of wife and mother.

Rita was closest to her beloved twin sister Paola, who she described as 'a part of herself'. Both girls were short, although Rita was the tallest, standing at five feet, three inches. They were not identical: Paola inherited their father's features and dark colouring, while Rita took after their mother and maternal grandmother, with large, limpid, grey-green eyes and paler colouring. Their interests were different too: Paola became a gifted artist whereas Rita was more inclined to academic pursuits.

Even though Rita was shy, school was by and large an enjoyable experience but not one, in the Italy of the 1920s, that encouraged girls to be anything other than wives and mothers. Science and mathematics teaching were non-existent in girls' high schools and only boys were expected to prepare for university and a career. Rita wasn't thinking about a career in science at that age but she had an innate sense that she would like to explore life beyond the restricted world of girls' education and Italian society's expectations. She later recalled, 'In my family, I saw that I could not ever accomplish anything, to be married like my mother was . . . I had no particular interest in children or in babies and I never remotely accepted my role as a wife or mother.'

Like many other countries, Italy suffered economically after the First World War ended and by 1920, when Rita was eleven, there were often workers' strikes and violent demonstrations. The time was ripe for a right-wing, authoritarian government and during the early 1920s, through a combination of propaganda and force, the Fascist party took hold. In 1922, Benito Mussolini came to power as prime minister, and ruled as dictator from 1925 to 1943 when he was finally overthrown.

In 1929, the family – still, as Rita put it, 'drifting along in the dark' – was struck by tragedy. Giovanna Bruttata, who lived with the Levis for a number of years and was like a second mother to the children, died of stomach cancer. At the age of twenty, Rita's academic ambitions had found a focus – she would train to be a doctor.

The first hurdle was to persuade her father. His sisters had earned doctorates in literature and mathematics but he blamed their unhappy marriages on their education. With her mother's input, however, Adamo swallowed his misgivings and eventually gave his approval. Rita began to cram furiously for her exams. For eight months, two local professors taught her languages, maths and science while she taught herself philosophy, literature and history. In autumn 1930, Rita was awarded a place to study medicine at Turin University, gaining the highest score of any of the home-taught students who took the entrance exams that year.

Rita was one of only seven women in their class of about three hundred students. After years of intellectual deprivation, she was in her element, declaring, 'I wanted to spend my time on research . . . I didn't want to spend any sentimental contact with other students, only intellectual contacts.'

Anatomy was Rita's favourite subject, fascinated as she was by the dissections of corpses in the medical school's Renaissance amphitheatre. Her professor, Giuseppe Levi, was not a relation but did remind Rita of her father, with his bear-like appearance and explosive personality. He devoted his career to studying cells, namely nerves, under the microscope, and Rita found this passion for histology magnetic. His rages were legendary, but his scientific influence was a powerful one. Three Nobel Prize winners, Rita and her lifelong friends, Salvador Luria and Renato Dulbecco, were trained in his laboratory. Renato Dulbecco identified the roots of their success in the 'right attitude for doing research' that Giuseppe Levi had instilled in them. He provided constructive, often strongly worded criticism when things had not gone to plan but equally became highly animated and excited when the results were interesting.

Rita's interest in the nervous system and the way it develops began at medical school. Neurons, with their central cell body and interconnecting fibres, are the functional cells of the nervous system and carry electrical impulses or 'messages'. These nerve cells produce a vast neural network in the brain and connect via the spinal cord to the peripheral nervous system throughout the body. Their specialised structure can only be seen under a microscope. The importance of understanding the development and function of

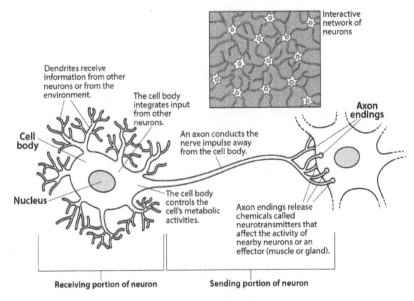

Interactive network of neurons

Dendrites receive information from other neurons or from the environment.

The cell body integrates input from other neurons.

Cell body

An axon conducts the nerve impulse away from the cell body.

Axon endings

Nucleus

The cell body controls the cell's metabolic activities.

Axon endings release chemicals called neurotransmitters that affect the activity of nearby neurons or an effector (muscle or gland).

Receiving portion of neuron   **Sending portion of neuron**

Structure of a neuron – the specialised, interactive and excitable nerve cell that receives, processes and transmits information throughout the body by electrical and chemical signals

neurons is highlighted by diseases such as Alzheimer's and Parkinson's disease, which are caused by damage to brain neurons. Symptoms of such diseases range from profound memory loss to lack of coordination, tremors and muscle rigidity.

In her second year at medical school, in 1932, Rita became Giuseppe Levi's intern. For her advanced histological studies, she had to cut very thin slices of nerve tissue and stain these with a silver-based dye (invented in 1873 by the Italian scientist Camillo Golgi), which preferentially stained neurons and revealed their intricate details. It was not an easy task but Rita rose to the challenge and became highly proficient at this sensitive technique, a crucial factor in her later success.

Giuseppe Levi was interested in finding out whether the number of neurons in the sensory ganglia (a cluster of intercon-nected neurons) on either side of the spine was the same or different in mice from different litters. Rita prepared endless microscope slides and counted thousands of cells. It was repetitive work and Rita doubted whether the counts were accurate or even relevant, later commenting that she found it 'more of an art than a science'. Although the necessary techniques and equipment were not available to carry this project further, Levi's research was ahead of his time. An important question remained and it was one that Rita would return to: were the numbers of nerve cells in particular parts of the nervous system fixed or altered by environmental influences?

When Rita was twenty-three, in her second year at medical school, her father became seriously ill, first with a stroke and then a series of heart attacks. At the beginning of August 1932 he died, aged sixty-five. Rita had not always had an easy relationship with him, but it was a devastating blow for the family and she joined them in mourning.

The following year, 1933, the Nazi party came to power in Germany, driving home its anti-Semitic policies from the offset; the first signs of Italian anti-Semitism appeared around this time. Hard at work, both in treating patients and on several projects in Giuseppe Levi's laboratory, Rita was largely immune to any impending threat. One of these projects, on the types of connective tissues (supports and binds other tissues of the body) that surround neurons, became her PhD thesis. Awarded in 1936, when Rita was twenty-seven, this propelled her further into a lifelong interest in nerve cell development.

As Mussolini's regime became more brutal, in the late 1930s, Rita worked with Fabio Visintini, examining the development of chick embryos. These are popular tools to study the development of the nervous system before birth. Chicken eggs are easy to obtain and an embryonic chick's nervous system is less complex to study than an adult bird or a mammal, but still highly informative.

It had been known since the early twentieth century that nerve cells pass electrical impulses to each other and Edgar Adrian, a

British physiologist, had made the first electrical recordings of this process in 1925. Using recently developed miniature electrodes, Fabio Visintini studied the electrical activity of the chick embryos' neurons and Rita's silver stain revealed the physical changes that occurred as the cells developed and differentiated. Together, they built up a comprehensive picture of the day-to-day development of a one to twenty-day-old chick's spinal cord and nervous system.

On 14 July 1938, 'The Manifesto of Race' was published in Italy and the anti-Semitic propaganda became harder to ignore, although Turin University, long known for its anti-Fascist stance, supported its Jewish staff. One member of the medical faculty, a friend since medical school, asked Rita to marry him on the grounds that this would protect her from anti-Jewish sentiment. Rita reminded him that she was not in the business of marrying anyone. Indeed, it soon became clear that such a match would be impossible. On 17 November 1938, the 'Laws for the Defence of the Race' forbade marriage between Aryans and other races. More devastatingly for Rita, it banned Jews from all teaching jobs and many other professions. Suddenly, most of Italy's 50,000 Jews were out of work, including Rita.

A job at the Neurologic Institute in Brussels, Belgium, provided only a brief reprieve, as six months later Italy entered the Second World War on the German side. Reluctant to stay away from Italy and her ageing mother, Rita returned to the family home in the hope that the war would be short-lived. It was a frustrating time. She continued to treat patients secretly but relied on the goodwill of her non-Jewish colleagues to write prescriptions for her, an increasingly difficult and dangerous practice. After a few months, Rita was forced to stop practising medicine and was left with little to occupy her except to read and visit friends.

In the autumn of 1940, when Rita was thirty-one, a casual conversation led to a new beginning. A former pupil of Giuseppe

Levi had recently returned from the USA and was dismayed to learn that Rita was not doing any research. Rodolfo Amprino's reaction ignited a spark in Rita. Inspired by an article she had read about the proposed mechanism of development of the nervous system in chick embryos, she set up her own home laboratory. As the Germans marched across Europe, Rita's bedroom became, as she described it, 'a private laboratory à la Robinson Crusoe'. There, she conducted the experiments that laid the foundations for her discovery of the first growth factor, molecules that influence the development of immature cells.

Her bedroom was not an ideal laboratory environment but, with the help of her architect brother Gino, Rita found ingenious ways of assembling the necessary equipment. The chicken eggs were incubated in a homemade incubator – a box with a thermostat. Ordinary sewing needles, sharpened on a fine grindstone, became surgical tools for operating on the tiny embryos. The local watchmaker was a source of miniature forceps and an ophthalmologist provided microscissors, more commonly used in eye surgery. Hidden away from the chaos of wartime and guarded by her mother, who dissuaded all intruders by declaring 'she is operating and cannot be disturbed', Rita began to make full use of her microsurgery and histology skills.

Rita's inspiration was the work of Viktor Hamburger, a German-born Jew who had fled to the United States. He was a founder of developmental neurobiology (the biology of the nervous system) who had made the chick embryo the standard research subject in this small but expanding field. Professor Hamburger had set out a long-term plan in 1927 to investigate how the nervous system developed.

He was particularly interested in one type of neuron, the motor neuron, which has its cell body in the spinal cord and extends its fibres out to the embryonic limbs. When this neuron is activated, it makes the muscles contract and the limbs move. In 1934, Professor Hamburger found that when the developing wing of a three-day-old embryo was cut off before the nerves reached it, the spinal ganglia, containing the neurons that would have served that wing, were much smaller than normal. Conversely, when extra limbs were grafted onto embryos, this resulted in a greater number of spinal motor neurons. He thought these results could be explained by a signal or inductive factor present in the wing.

Six years later, in her tiny hidden bedroom laboratory, Rita was keen to repeat these experiments and see if she came to the same conclusion. In an interesting role reversal, she was often helped by Giuseppe Levi. This was a mixed blessing, as his size and clumsiness led to a number of breakages, but she welcomed his advice.

After removing limb buds from three-day-old embryos, she dissected a few embryos every six hours for the next seventeen days. Microscopic slides, with sections from these embryos, were silver stained and the growth of the neurons from the nearby ganglia was revealed. Rita was fascinated to find that the neurons did continue to grow towards the stump of the amputated wing but then they died once they reached the stump. It appeared that an unknown factor in the wing bud attracted the nerve cells growing from the spinal cord ganglia, encouraging their growth but not their survival or further development. Without this unidentified factor, the neurons died.

Professor Hamburger had hypothesised that the limb buds contained a substance that directs immature nerve cells to differentiate into motor neurons, an inductive influence (trigger of nerve growth) belonging to the category of those known as 'organisers'. Rita interpreted the data differently, suggesting that the substance in the limb buds promoted the survival of the newborn nerve cells.

The interpretation of Rita and Professor Hamburger's results was subtly different, but Rita's intuition convinced her that she was correct. This was the first indication of an important trophic (nutritional) factor but also led to a new concept – that nerve cell death is a normal part of development. Rita realised, 'it was a pure miracle that I succeeded with such primitive instrumentation', and, like others, later marvelled at her ability to focus so intently on her research while war raged round her. She felt that 'the answer lay in the desperate and partially unconscious desire of human beings to ignore what is happening in situations where full awareness might lead one to self-destruction'.

Rita's findings were initially published in a brief paper in a Belgian journal, *Archives de Biologie*, because no Italian journal would publish research by a Jew. Soon after, in 1942, the peaceful world of her bedroom was shattered when the industrial town of Turin became a key target for British bombers. The family sheltered in the basement, with Rita clutching her precious microscopes and slides. Facing up to the inevitable destruction, Rita and the family escaped to the hills to a town called Asti, where Rita carried on with her research.

There, the conditions were even more cramped than before: Rita's 'laboratory' consisted of a small table in the corner of the living room, where her work was hampered by regular power cuts and a shortage of eggs. Rita, ever resourceful, cycled round the local farms asking for eggs for 'her babies', explaining that the fertilised ones from the roosters were 'more nutritious'. Much to her brother Gino's disgust, after extracting the embryos for her research, she made omelettes from the eggs, thus supplementing the family's wartime diet.

The conditions were challenging but Rita was undaunted, saying later that her analysis of the nervous system 'came into focus and grew in that country milieu, probably much better than it would have in an academic institution'. Rita was fascinated by a new project – the study of nerves in the chick embryo's developing ear. She closely observed the passage of nerve fibres leaving their clusters of cells near the embryo's spine and migrating along the same path to their destination in the chick's body. She likened it to ducklings followed their mother. It was a predictable process with some cells dying, as an entirely normal part of a healthy embryo's development, and reminded Rita of the cycles of life and death she saw in the countryside around her.

In the relative seclusion of rural Italy, Rita and the family were protected from the war, which was not going well for Mussolini. In April 1941, the Allies had pushed the Italians out of north-east Africa and in July 1943 they invaded Italy; Mussolini was forced to resign and was duly arrested but the Levi family's rejoicing at the news was short-lived. In September 1943, German troops advanced into Italy, unopposed by Italy's diminished army. As Rita wrote in her autobiography, it suddenly become apparent that 'a delay of days, perhaps even of hours, might cost us our lives'. The Levis avoided deportation to the concentration camps, travelling by train to Florence with fake identity papers that Rita and her sister Paola had made. For nearly two years, they lived under assumed names with a Catholic identity, in constant fear of discovery.

Rita was forced to abandon her precious home laboratory and all its contents when they fled. She and Paola occupied their time by making more false identity papers and listening anxiously to the news broadcasts from London. In spring 1944, some respite was provided by the welcome arrival of Giuseppe Levi and Rita spent the next few months helping him to revise his histology textbook.

On 3 August 1944, a state of emergency was declared and the city's electricity and water were switched off. As the Italians fought the German occupation, Rita and her family were frightened by the scenes of destruction they witnessed from their apartment but Rita also felt 'an exhilarating air of freedom'. On 2 September 1944, British troops seized control of Florence and Rita immediately found a new sense of purpose. For the next few months, until May 1945, she worked as a doctor in the refugee camps. As she wrote in her autobiography, this was her 'most intense, most exhausting and final experience as a medical doctor'. Death stalked the camps, particularly for the young babies that arrived weakened by cold and hunger, and, with no antibiotics available, the death toll from infectious diseases like typhoid was high.

Rita survived this sober and worrying time to greet the advent of a new age for Italy. On 25 April 1945, the Allies and their partisan supporters drove the last of the Germans out of Italy. The Levi family and Giuseppe Levi returned to Turin to try and pick up the pieces of their life there. Rita, not surprisingly, found it hard to settle and adapt. But, in July 1946, her spirts were immediately lifted by a letter from Viktor Hamburger, who had read about Rita's duplication of his experiments in the makeshift conditions of her home laboratory, and who was now based at Washington University, St Louis.

Rita recognised it was a great honour to be invited to work with such an eminent scientist. At the age of thirty-seven, she was invited for a one-semester visit. Her father's opposition to female education combined with the chauvinist nature of Italian society, anti-Semitism and the Second World War had meant a slow start to Rita's career in science. Little did she know that this visit would extend to nearly thirty years. Rita later recalled that her years at Washington University were the 'happiest and most productive years of my life'.

Rita was excited about the prospect of starting a new chapter in St Louis, which had a reputation as a frontier city. Washington University was founded in 1853 and its ivy-covered brick buildings and peaceful atmosphere had a calming effect on Rita. She enjoyed exploring her new surroundings, but she found her first two years in St Louis hard; her English was poor and she was initially bemused by the informal nature of American universities. She was immediately taken with Viktor Hamburger, however, and, like many others, found him to be kind and generous with a dignified manner. His incremental approach to scientific research complemented Rita's more flamboyant Italian style and they became close working colleagues.

Viktor was intrigued by the possibility of combining his research interest – embryology – with Rita's focus on the nervous system – neurobiology. The German scientists, Hilde Mangold and Hans Spemann, were instrumental in setting Viktor on his career path. In 1924, they had shown that a region of the early (gastrula stage) amphibian embryo acted as an 'organiser', instructing the formation of different parts of the embryo. This was the first demonstration of the concept of 'induction' – the interaction between two groups of cells, such that one group influences the developmental fate of the other.

Viktor and Rita were both interested in how the complex network of neurons innervate peripheral tissue – how nerves grow from the spinal cord into the limbs. Viktor had been Hans Spemann's student at the time of the 'organiser' discovery and thought that 'induction' might be responsible for the nerve growth seen in his and Rita's chick-wing experiments.

In the late 1940s at Washington University, Rita had the opportunity to repeat some of her and Viktor's experiments, where they removed or grafted limbs onto chick embryos and looked at the resulting dramatic effect on nerve growth. Her results persuaded Viktor that her interpretation of the results might be correct: a factor in the limb was promoting the survival of the newborn nerve cells rather than directing immature nerve cells to differentiate into motor neurons. Viktor was so impressed with Rita's skill and dedication that in 1947 he extended her short-term contract and offered her the position of research associate.

That autumn, Rita's histology skills, combined with her powers of observation yielded more results. Rita was studying slides made from a chick embryo between day three and day seven of development, when neurons in the bird's brain and spinal cords were beginning to form. She noticed that rapid changes were taking place in certain areas of the spinal cord, often over a period of just a few hours. Like a 'large army on a battlefield', she observed long lines of

neurons streaming from one place to another. This migratory movement seemed to Rita to be almost intuitive, like migratory birds
responding to a 'programme' inside them. Rita's highly developed
intuition was instrumental in the success of such experiments and
their interpretation, something of which she was well aware: 'I have
no particular intelligence, just average intelligence. But, intuition is
something that comes to my mind, and I know it's true. It's a particular gift in the subconscious.'

In another area of the slides, she saw dying cells, resembling
'corpses' on her battlefield and, later on, these cells were 'eaten' by
macrophages, the scavenger cells of the immune system. It was a
dynamic, constantly changing system that highlighted the cycle of
growth, cell migration and death that she had first seen in Asti in the
spring of 1942. She had found direct evidence that large numbers of
neurons die during the course of their normal development and that
amputating an embryonic limb makes even more of them die. The
life of a developing neuron depends on some kind of signal from the
limbs; without it, they die.

In January 1950, Rita's research made a leap forward. With the
help of Elmer Bueker, a former pupil of Viktor Hamburger, she
showed that a cancerous mouse tumour transplanted onto a three-
day-old chick embryo stimulated nerve growth from the embryo's
sensory ganglia into the tumour cells. The nerve growth reminded
Rita of 'rivulets of water flowing steadily over a bed of stones'. Other
mouse tumours produced a similar effect, with such vigorous nerve
growth that Rita thought she was hallucinating.

Now Rita made a critical observation. The nerve fibres were
present inside the embryo's veins but not the arteries. Veins would
carry substances from the tumour to the embryo whereas the reverse
was true for the arteries. The tumours were clearly producing a fluid
substance that stimulated nerve growth. Rita devised an ingenious
new method to confirm and extend her findings. Instead of attach-
ing a tumour directly to the surface of the embryo, she grafted the

tumour to the protective outer membrane. The experimental adaptation prevented direct contact between the tumour and the embryo but allowed the passage of any active substance, contained within the tumour cells, across the semipermeable membrane to the embryo.

To her delight, Rita observed the same results as she had seen before. The prodigious nerve growth was stimulated by a diffusible chemical produced by the tumour cells. In January 1951, she corresponded with Renato Dulbecco, who described her findings as 'sensational', and Paul Weiss, the eminent embryologist, who heard Rita talk at the New York Academy of Sciences, thought it was the most exciting discovery of the year.

Rita's work in St Louis continued unabated until she went to Brazil to learn a new, simpler and more rapid technique that wasn't available in Viktor Hamburger's laboratory, but one that would significantly advance her research. Armed with a grant from the Rockefeller Foundation, Rita flew to Rio de Janeiro in September 1952. Rita had two other vital weapons under her belt, quite literally – a pair of mice carrying the tumours she wanted to test, hidden in her hand luggage. The next few months unfolded in a highly productive period of research for Rita and she detailed her findings to Viktor in their regular correspondence. Rita faced a daily challenge to identify the substance that promoted nerve growth and Viktor was struck by the way her letters illustrated 'beautifully how real research works, ups and downs, despair and triumph'.

Growing nerve cells outside an organism *in vitro*, meaning in glass, had only been possible for about twenty years and cell culture was not commonly used in the 1950s, but Rita realised that this was the way forward. *In vitro* cell culture provided the ideal environment to control and investigate factors that influence the growth, development and properties of neurons. A former colleague of hers, Hertha Mayer, who had worked with Rita in the 1930s, was now head of a tissue culture laboratory at the Institute of Biophysics in Rio de

Janeiro, Brazil. Rita was keen to learn the new techniques that were being practised there.

Initially Rita found it difficult to reproduce her results in a tissue culture system but, with some adaptations to the model, she clearly showed that fragments of tumour cells could initiate nerve growth. Normal mouse tissue also made nerve fibres grow, albeit not to the same extent. Rita had originally thought that only tumour cells, which grow rapidly and uncontrolled, could produce this growth-promoting substance. Now she wondered whether it was a property of mouse tissue in general.

As her time in Brazil came to an end, Rita spent her last month in Rio as a tourist: a rare opportunity to rest and relax. She missed the famous carnival itself, but attended a pre-carnival event, which left a strong image on her mind. Rita was beginning to see her elusive molecule as a mysterious masked figure, like those she saw on Rio's crowded streets. In her autobiography she wrote, 'It was in Rio de Janeiro that it revealed itself . . . in a theatrical and grandiose way, as if spurred by the bright atmosphere of that explosive and exuberant manifestation of life that is the carnival in Rio.'

Buoyed by her Brazilian experience, Rita was all set to isolate and identify the factor responsible for the nerve growth, a project she thought would only take a few months but which, in reality, took six years. In 1953, aged forty-four, Rita returned to Viktor Hamburger's laboratory at Washington University in St Louis, where she was delighted to find that Stanley Cohen was joining the laboratory. Stanley was a Brooklyn boy, down to earth, yet modest and reserved, in sharp contrast to the Italian Rita, famous for her exquisite dinner parties, conversant in several languages and, with an artist twin, fascinated by the arts. For the second time in her working career, Rita found herself in a complementary and close working partnership. Stanley, as a biochemist, knew nothing about the nervous system and, equally, Rita was no expert in biochemistry.

Their approach to science was quite different too: Stanley was meticulous in his consistent, cover-all-bases approach whereas Rita followed her famous intuition on a regular basis, often with spectacular results. He admired her 'great drive to succeed' and they complemented each other well, with Stanley once saying, 'Rita, you and I are good, but together we are wonderful.'

Like Rita, Stanley had been fascinated since his undergraduate days by the embryonic development of cells and had come to Washington University in 1952 to learn how to use radioactive materials in biological research. He embarked on a mission to work out the chemical nature of the substance that Rita had discovered. In 1954, Rita and Stanley named this substance nerve growth factor (NGF).

Rita set about producing a large enough sample for Stanley to test, extracting NGF from dozens of tumours grown on embryos. After several years of hard work, this source enabled Stanley to determine that NGF was a protein. And they discovered a highly concentrated source of NGF – 1000 times more concentrated than in tumour cells – in the form of snake venom. This greatly facilitated their subsequent biochemical analysis but also led to the finding that most mouse tissues produced some NGF, with varying levels of production.

Over the next three years, 1956–9, Stanley discovered that the molecular weight (the sum of the atomic weights of all the atoms in a molecule, measured in daltons) of NGF was 20,000 Da, while Rita concentrated on its biological effects *in vivo*. For example, she showed that a daily NGF injection over three weeks stimulated a tenfold increase in the size of the nerve ganglia in newborn mice and rats, and an increase in nerve growth into the surrounding organs and tissues.

Most importantly, Rita demonstrated that this effect was specific by blocking it. In biological systems, this can be achieved with an antibody. These are routinely produced by the immune system in response to a foreign insult – usually a microorganism such as a bacteria and virus – allowing a specific and targeted immune response to occur. When Rita used an antiserum (containing anti-bodies against mouse NGF) *in vitro*, there was no growth of nerve fibre halos in her tissue cultures system.

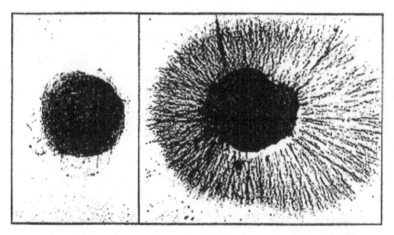

Chick embryo nerve growth in the absence (left) and presence (right) of NGF
Levi-Montalcini, R and Calissano. The Nerve-Growth Factor. Scientific American. 1979. Vol 240, pp 68-77

In vivo, the effects were even more marked. After a month of daily antiserum injections, the ganglia along the spinal cords of newborn mice and rats were almost completely obliterated. The rest of the rodents' nerve systems, however, developed normally and the mice appeared healthy. Viktor Hamburger visited the day they made this discovery and remarked that 11 June 1959 was a 'memorable event in neuroembryology'. NGF not only stimulated nerve growth but certain types of nerves depended on NGF for their survival and development. This seminal work was published in 1960 in the *Proceedings of the National Academy of Sciences*, when Rita was fifty-one years old.

After six productive years, at the end of 1958, Viktor found he couldn't afford to offer Stanley Cohen a permanent position as he was a biochemist but would be required to teach zoology, and he subsequently found another job, at Vanderbilt University in Tennessee. Nowadays, science is interdisciplinary and a zoology department could routinely employ biochemists but the practice was uncommon in the 1950s. It was a blow to Rita and she heard the news, as she said later, 'like the tolling of a funeral bell'. But, with her appointment as a full professor in 1958, and the arrival of an Italian biochemist called Pietro Angeletti in 1959, Rita's work continued uninterrupted.

As Rita approached her fifteenth year at Washington University, aged fifty-two, she decided that she would live in Italy for part of the year, while Pietro Angeletti ran the Washington laboratory. The Levi family had been regular correspondents and Rita visited for a month each summer, but she missed seeing them more frequently and her mother was now quite elderly. With dual citizenship, she was free to reside in both countries and decided she would work in both, sharing an apartment when she was in Rome with her twin sister Paola.

The Italian government was not renowned for its support of science in the 1960s but Rita's growing reputation helped her, and she was able to find laboratory space and equipment from the country's Institute of Health and raise some research funds from the National Research Fund. Rita had no trouble recruiting either, as young scientists were keen to work for her, even on a meagre salary. In 1961, when Pietro Angeletti returned to Italy, he and Rita established a new Centre of Neurobiology in Rome's Advanced Institute of Health.

It took some adjusting to life back in Italy. Just as she had initially found the informality of Washington University hard to get used to, now the reverse was true and she was struck by the formal nature of the European students' relationships with their professors. Characteristically, Rita rose to the challenge. As a young woman at university, Rita had dressed plainly, 'like a nun', but by her fifties, immersed again in Italian society, she was elegantly turned out, with beautiful dresses and matching silk or brocade jackets tailor-made for her. She was short in stature but even when her attire was covered by a laboratory coat, her styled hair, carefully chosen matching jewellery and commanding presence led one of her former students to name her the 'Queen of Italy in a lab coat'.

It was the beginning of an unsettled period for Rita. She found it difficult to balance her commitments to her Italian family and American friends; in 1963, she lost her mother and, two years later, her mentor and old friend Giuseppe Levi. In 1969, Rita was made director of the new Laboratory for Cell Biology in Rome; the transition was not an entirely smooth process, particularly as Rita now felt her research wasn't attracting enough interest.

NGF was a completely novel biological phenomena and neurologists found it hard to accept its existence. There is some debate as to whether she was correct in her perception, particularly as she had a reputation of being possessive about her work. Rita was always strong and forthright in her views; her determined nature no doubt helped her progress in science but did not always endear her to others. Suffice to say she felt strongly enough to turn her back on her first love and spent several years working on the nervous system of cockroaches.

In 1972, Rita was drawn back to her original line of work and her American and Italian laboratories became more focused than ever on understanding NGF. Pietro Angeletti and a chemist, Vincenzo Bocchini, succeeded in making pure enough samples of NGF to obtain its specific sequence of amino acids – the twenty or so building blocks that make up all proteins. NGF is comprised of two identical amino acid chains, each 118 amino acids long.

The interactions between NGF and other molecules, particular receptors, were also attracting interest. Signalling molecules, such as growth factors, hormones and immunoregulatory molecules, need to latch onto cells using protein molecules called receptors.

Each signalling molecule has its own receptor, and the first receptor for NGF was identified in the early 1970s, with a second one twenty years later.

In the mid-1960s, Stanley Cohen had found another type of growth factor, epidermal growth factor (EGF), which made the outer layer or epidermis of the skin grow faster than normal. Stanley subsequently showed that when growth factors, like NGF and EGF, bind to their receptors, the paired molecules of the receptor, enclosing the growth factor, are taken inside the cells. Here, the growth factor sets to work initiating growth, differentiation or even death, depending on the precise function of the growth factor in that cell type.

Working out the mechanism of NGF's action took a number of years but, by 1974, Rita was delighted to find that the original mechanism that she had proposed was correct. Nerve growth factor stops nerve cells from dying in their early embryonic stage. Without NGF, up to half these cells will die. Tissues in the developing embryo innervate (become supplied with nerves) along a concentration gradient of NGF. For example, the embryonic limbs that Rita dissected in her earliest experiments produced a high concentration of NGF compared to the ganglia near the spinal cord. NGF is taken up by the ends of sympathetic (responsible for the body's 'fight or flight' responses) or sensory nerve fibres and is transported back along these fibre to the cell body, thereby stimulating growth of the nerves towards the limbs or other tissues.

Nerve growth factor initially interested neurobiologists because of its effects in the developing nervous system. Throughout the 1980s, the research community's burgeoning interest in NGF, and its interactions and effects, was reflected in the number of papers published – well over a thousand in that decade alone. It is now clear that NGF functions throughout the life of the animal with a wide repertoire of actions, including effects on the production of hormones and modulation of the immune system.

It was research into NGF and its influence on the immune system that gave the first hints of a mechanism through which the nervous and immune systems might interact. In addition, some cancer-causing genes (oncogenes) were shown to be mutated versions of growth factor genes, leading to the uncontrolled growth seen in cancer cells. These key findings, and many others, helped push NGF and other growth factors further into the limelight.

The essential and ubiquitous nature of NGF's actions was further highlighted in 1983 when its gene sequence was discovered. The gene is highly conserved, suggesting that the protein for which it encodes performs vital functions in vertebrates. Furthermore, the NGF gene is located on the short arm of chromosome 1, in a region that was more easily damaged or deleted than other regions, another finding that had implications for NGF's role in disease.

After Rita's official retirement in 1977, at the age of sixty-eight, she carried on working at Washington University in an unofficial capacity, saying that 'when you stop working, you are dead'. Since Rita's discovery of NGF, numerous other growth factors that stimulate cell growth, differentiation, survival, inflammation and tissue repair throughout the body have been discovered. The list now extends to well over a hundred, grouped into more than ten families. Many of these are being investigated or are currently in therapeutic use. For example, the epidermal growth factor (EGF), discovered by Stanley Cohen, is now used routinely in wound healing for burns victims and corneal transplants.

Rita and Stanley had opened the world's eyes to the existence and power of growth factors and, on 13 October 1986, they received recognition for their lifelong work. That year's recipients of the Nobel Prize in Physiology or Medicine were Rita Levi-Montalcini and Stanley Cohen for their 'discoveries of growth factors', NGF and EGF respectively. Rita, at the age of seventy-seven, was the fourth woman to win this Nobel Prize and the first Italian woman to win any scientific Nobel Prize. As the Nobel Assembly noted, Rita's discovery of NGF 'is a fascinating example of how a skilled observer can create a concept out of apparent chaos'.

The Nobel Prize was only one of the numerous awards Rita won. For example, in 1987 she received the National Medal of Science, the highest scientific award given by the US government, and she was the first woman to be a member of Italy's Pontifical Academy of Sciences. Rita unashamedly used her increasing fame to promote the cause of Italian science, so much so that the Italians even began to joke that Pope Jean Paul II was instantly recognisable, provided he appeared alongside Rita Levi-Montalcini.

Rita's research methods and working environment may, at times, have been unorthodox but she produced results and her written output was prodigious. As well as over two hundred scientific papers, she wrote several books including her autobiography, *In Praise of Imperfection*, published in 1988, and an anthology of key scientific papers entitled *The Saga of the Nerve Growth Factor*, published in 1997; both the autobiography and the anthology are translated into English.

The reviews of Rita's autobiography were mixed, with some suggesting she had presented a fairy-tale view of the way science is conducted and hadn't recognised the contribution of her fellow scientists sufficiently. Many, including Rita herself, regarded her as the mother of growth factors and like all mothers she was fiercely possessive and protective of her 'child'. Rita was certainly irritated by the time it took for the scientific community to recognise the potential importance of growth factors. Stanley Cohen was more sanguine, saying that 'this had the advantage that people left you alone and you weren't competing with the world. The disadvantage was that you had to convince people that what you were working with was real.'

Rita was realistic about her rollercoaster life and personality – she was once described as a cross between Marie Curie and Maria Callas – and took this into account when she entitled the book *In Pursuit of*

Imperfection. This refers to the Irish poet, William Butler Yeats, who thought that the perfection of life and the perfection of work were irreconcilable. Rita begged to differ: 'The fact that the activities that I have carried out in such imperfect ways have been and still are for me a source of inexhaustible joy, leads me to believe that imperfection rather than perfection ... is more in keeping with human nature.'

As she got older, Rita increasingly recognised the challenges faced by young people, particularly women in society, and applied her still considerable energies to charitable work. In 1992, when Rita was eighty-three, she and her twin sister Paola set up a charitable foundation, funded with Rita's life savings and in memory of their father. Rita had been struck by the similarities between the discrimination against African girls, who are often denied access to education, and her experiences as a Jew and a woman in Fascist Italy in the 1930s and 1940s. Rita felt strongly that women should be empowered, saying, 'After centuries of dormancy, young women can look forward to a future moulded by their own hands.' The charity has funded hundreds of girls through their studies, including a number of medical students. Its website stresses the importance of giving girls the tools to develop their full potential and the influence this has on the 'economic, social and cultural development of society as a whole'.

Rita also helped women in other ways. With the sociologist Eleonora Barbieri Masini, and others, she set up the Women's International Network, Emergency and Solidarity (WIN) in 1995. This publishes a directory of organisations that focus on providing assistance for women affected by poverty, prostitution, drug abuse, religious conflicts, migration and a whole range of current issues.

Like her scientific work, Rita was never afraid to tackle problems head on, even if that resulted in heated discussions. After Rita screamed at somebody, her former secretary and friend Martha Fuermann commented, 'Her temper doesn't last. She doesn't hold a grudge. Two things are important to her: her work and her twin, almost in that order.'

By the time she was in her nineties, Rita's failing eyesight prevented her doing research at the bench, but she still juggled her various scientific and other commitments. As a member of the Italian Senate, Rita attended meetings there into her late nineties, when she was the eldest present. Her long-standing involvement in and contribution to Italian science meant that she was frequently called upon to give policy speeches on, for example, the effects of the latest tax plan on science.

Rita's contribution to the world of science, and growth factors in particular, was obvious long before her death in 2012. NGF has been described as a 'Rosetta Stone', helping to decode key aspects of nervous system function, proteins, receptors and cancer biology. As the first growth factor identified, NGF introduced a new concept by suggesting that there were another means of communication between cells, at short range, rather than via blood-borne substances such as insulin. Growth factors, and the interactions they facilitate between the different organ systems in the body, have helped change scientists' view of the way the nervous system operates, not just as an organ that monitors and controls the body but one that is controlled by the body.

Over the decades since NGF was discovered, it has been shown to be a key link between the nervous system, the endocrine system (hormones) and the immune system. Connections between these systems had been suggested as early as the beginning of the twentieth century but Rita was the first to propose that NGF was involved.

For example, the 'fight or flight' response, which involves the sympathetic nerves and is the body's response to stressful situations, is regulated by NGF. Starting in the nervous system, this response spreads to the endocrine (hormone) system and the immune system. Rita first demonstrated this connection between NGF and stress in the late 1980s, when she was studying fighting mice. NGF levels increased not only in the salivary glands but also in the bloodstream and their hypothalamus. This area of the brain signals to the nerves that control blood pressure and heart rate and also triggers the pituitary gland hormone cascade, resulting in the production of stress hormones and downstream effects on the immune system. Cells in the pituitary gland, the brain and a key cell of the immune system, the mast cell, all carry receptors for NGF.

Lifelong learning is also associated with NGF. Over the past twenty years or so, scientists have realised that the brain retains some plasticity even into old age. Contrary to previous ideas, it's now known that new connections can be made, even after the damage caused by, for example, head injuries. This had been suspected since 1989 when Rita and other scientists demonstrated that NGF stimulated new fibres to grow from the ends of brain neurons in adult rodents.

These findings were given further support in 2001 by the neurologist Howard Federoff's group. Mouse brain nerves cells in the hippocampus (the area of the brain associated with memory, particularly long-term memory, and spatial recognition) were genetically modified to produce larger quantities of NGF. These mice, in comparison to normal controls, had a greater number of neurons connecting the different areas of the brain involved in learning, and were much quicker at learning new tasks, for example negotiating complex mazes.

In 2002, Rita founded the European Brain Research Institute in Rome. It was set up to study the nervous system from its development to its role in health and diseases. Finding new therapies for neurodegenerative diseases, such as Alzheimer's, Parkinson's and Huntington's, is a particular focus. Nerve growth factor belongs to a family of factors called neurotrophins that may help to slow the degeneration of the nervous system in these diseases. NGF is active not only in the peripheral nervous system but also in the brain. All three of these diseases affect neurons that have been shown to produce NGF and/or to respond to it.

In 2015, Mark Tuszynski and his colleagues published, in the journal *JAMA Neurology*, the ten-year follow-up to his landmark study of experimental gene therapy in Alzheimer's patients. Targeted injection of the NGF gene into the brains of these patients rescued dying cells around the injection site and led to an enhanced growth of these neurons. In some patients, this effect was still evident ten years after the therapy was administered. This was only a small Phase I trial to test for safety but the preliminary results from the Phase II trial, testing for efficacy, suggest that mental function declines more slowly in the patients involved.

As research reveals the biology of NGF, its role as a regulator of growth, namely in cancer cells, has been highlighted. Signalling of NGF through its receptors has been shown to alter death and survival in various cancer cells and is attracting much interest as a target for cancer treatments. For example, in an animal model of breast cancer, antibodies against NGF were successful in reducing tumour growth, suggesting that anti-NGF antibody therapies may prove useful in the treatment of breast cancer where overexpression of NGF is a factor. And anti-NGF antibodies have been shown to reduce cell migration by up to 40 per cent in two prostate cancer cell lines *in vitro*, revealing the potential to limit cancer metastasis.

In the early 1990s, Rita's team showed that NGF is involved in the pain of inflammation by making nerve endings in the inflamed part of the body more sensitive to pressure and heat. This finding opened up a new field of painkilling drug research. For example, one drug, Tanezumab® (a monoclonal antibody to NGF), reached Phase III clinical trials, proving highly effective at blocking pain perception in patients with chronic pain from osteoarthritis of the knee. Although the trial was prematurely halted in June 2010 because some participants presented with increased damage to joints, it's likely that the reason for these adverse effects was excessive use of the joints as a result of lack of pain sensation.

With many of these potential new therapies, there are challenges to be overcome. Unsurprisingly, for such a wide-acting and powerful molecule, deleterious side effects are common and maintaining a pharmacologically active dose can be problematic. The list of questions about NGF's biology, particularly those that relate to its use as a therapy, is still a long one. For example, using the X-ray crystallography techniques pioneered by Dorothy Hodgkin (see Chapter Five), scientists can study the 3D structure formed when NGF, in naturally occurring or engineered therapeutic forms, interacts with its receptors.

In 2009, aged 100, Rita proposed her own agenda for NGF-related research. In addition to work on its therapeutic potential, she suggested an investigation into a potential role for NGF in reproduction and human fertility. As ever, her scientific intuition was proved correct. Three years later, in 2012, a paper was published on a substance, discovered in the semen of a diverse range of mammals, that induces ovulation. It was NGF.

As Rita passed into her second century, she shared her secret of longevity. She thought it was related, in part, to a high metabolism induced by eating 10 per cent less calories than normal and eating just one meal a day. 'I might allow myself a bowl of soup or an orange in the evening, but that's about it,' she said, 'I'm not really interested in food, or sleep.' Work and keeping her brain active were highest on her agenda. In 2009, she said in an interview, 'At 100, I have a mind that is superior – thanks to experience – than when I was 20.' Rising at 5 a.m., she would pass on that knowledge, spending the mornings in the laboratory directing her all-female team and the afternoons at her African foundation.

Rita outlived her family too: her brother Gino died in 1974 aged seventy-two, her twin Paola in 2000 aged ninety-one, and her other sister Anna not long afterwards. Rita Levi-Montalcini's long life ended on 30 December 2012, at the age of 103, at home in Rome. Her funeral was held at the Italian Senate and her body was cremated in Turin the following day. Rita's ashes are buried in the family tomb at the Hebrew Monumental Cemetery in Turin.

Rita's association with NGF continued until the year she died, when the last paper bearing her name was published. Printed in the American journal *Proceedings of the National Academy of Sciences* (PNAS), it was entitled, 'Nerve growth factor regulates axial rotation during early stages of chick embryo development'. Over sixty years after the first growth factor, NGF, was discovered, scientists are still finding out new facts about this crucial molecule and often in the same model system that Rita was using in her hidden laboratory in 1930s Fascist Italy.

Lise Meitner
(1878–1968)

In the early twentieth century, it was barely possible for women to work in science at all and yet two of them were at the forefront of understanding the structure of the atom. The first was one of the few female scientists of which people have heard, the Polish scientist Marie Curie (see Chapter 3), who pioneered the study of radioactivity. The diminutive Austrian Lise Meitner, who Albert Einstein once called Germany's own Marie Curie, was the other.

One of the most brilliant nuclear scientists working in Germany, Lise fled Hitler's regime when it was almost too late. In exile, she made the startling discovery that a nucleus of a uranium atom could be split in two, releasing phenomenal amounts of energy – a discovery that would immediately revolutionise nuclear physics and lead to the atomic bomb. While Lise was recognised after the Second World War as the 'mother of the atomic bomb', she wanted, and indeed had, no part in its construction and her true scientific contribution became obscured in subsequent years. In 1944, the Nobel Prize in Chemistry was awarded to Lise's collaborator Otto Hahn for this discovery, overlooking her involvement.

Lise Meitner was born on 7 November 1878 in Vienna, then the capital of the Austro-Hungarian Empire. She was the third of eight children, five girls and three boys, in an intellectual Jewish family. Elise was her birth name but she would later shorten this to Lise. Her father, Philipp Meitner, was one of the first Jewish lawyers in Vienna, a man who was socially and politically active in spite of a climate of anti-Semitism.

Lise grew up in Leopoldstadt, a district of Vienna that was predominantly Jewish. Despite this, the children grew up in a largely agnostic household and, in 1908, Lise's two sisters became Catholics and she herself became a Protestant. Lise was one of the many Jews who later discovered that the Nazis did not count Christian baptism as a mitigating factor in their genocide project.

Philipp was a freethinker and their home was a focal point for all kinds of people to visit: lawyers, writers, chess players, intellectuals and politicians. The children were allowed to stay up late and listen to the adults expound their views. Lise later reflected on 'the unusual goodness of my parents, and the extraordinary stimulating atmosphere in which my brothers and sisters and I grew up'.

Hedvig Meitner was a strong maternal presence. As a talented pianist, she gave the Meitner children music lessons and taught them to be independent, saying, 'Listen to me and your father but think for yourself!' The children were encouraged to study science and, even as an eight-year-old, Lise's inclinations were clear. In her notebook, she recorded how the thin coatings of oil films reflect light. With her cultured upbringing, music and physics became Lise's lifelong passions.

From an early age, Lise was a natural sceptic. She challenged and examined the world around her, often using reason and critical thinking as a weapon against superstition. When she was a little girl, her grandmother warned her never to sew on the Sabbath, otherwise the heavens would fall. As Lise tentatively inserted her needle tip into her embroidery, she glanced upwards. With no obvious thunderclaps, she made another stitch, now satisfied that her grandmother's words were inaccurate.

Lise's determination to learn meant that, by the time she was eight years old, she was hiding maths books under her pillow and blocking the crack beneath her bedroom door so she could study past bedtime undetected. Lise's parents encouraged all their children to gain an education, although that was a challenge for the three girls, when Austrian society only expected them to learn enough to run a household efficiently. Much to her frustration, Lise's schooling in Vienna ended when she was fourteen, as there were strict restrictions on female education.

In 1867, Austrian universities were opened to men from any background but not to women. A small number of women had asked professors whether they could attend lectures but the most they were allowed to do was to sit in on classes, never in an official capacity, and there were no academic awards given. As a teenager, Lise dreamt of becoming a physicist, an impractical aim in many ways. There were very few jobs in industry for physicists and it was generally considered a dead subject on the grounds there was very little left to be learnt about the physical world, apart from confirming a few measurements!

Lise called the years from 1892 to 1901 her 'lost years' but when the restrictions were lifted, in 1901, she obtained her high school certificate, aged twenty-two. She compressed eight years' worth of study in logic, literature, Greek, Latin, botany, zoology, physics and mathematics into twenty months. Whenever she took a break, her brothers and sisters teased her, saying, 'Lise, you're going to flunk. You have just walked through the room without studying', and photographs of the time show a pale young woman with dark rings under her eyes. Although Lise demonstrated a gift for mathematics from an early age, there was little correlation between aptitude and opportunity for women in nineteenth-century Europe. Lise later remarked, 'Thinking back . . . to the time of my youth, one realises with astonishment how many problems then existed in the lives of ordinary young girls, which now seem unimaginable. Among the most difficult of these problems was the possibility of normal intellectual training.'

With the support of her parents, Lise took the entrance exam to university with thirteen other students. Only Lise and three others passed and, in October 1901, at nearly twenty-three years old, Lise started at the University of Vienna. All eight of the Meitner children were encouraged – both boys and girls – to purse their passions and study for advanced degrees. Lise's youngest sister Frida earned a physics PhD and became a college professor while her older sister became a concert pianist. One of her brothers, Walter, was awarded a chemistry PhD and another, Fritz, became an engineer.

Lise was the first woman admitted to the University of Vienna's physics department. Female students were generally thought of as misfits and it didn't help that Lise was painfully shy. In her free time, she went to concerts, often alone. Buying the cheapest seats at the Vienna Court Opera, she sat high up in the area she called her 'musical heaven' and followed the music with her own scores.

At university, Lise had a number of inspiring lecturers. Professor Frank Exner, who made the university an early centre for radioactivity research, brought such clarity and perspective to physics that students from all disciplines came to hear his lectures. In her second year, she joined a small group of like-minded women to study with the theoretical physicist Ludwig Boltzmann. He was an 'atomist', famous for developing kinetic theory (the theory that the minute particles of all matter are in constant motion) and the statistical analysis of the motion of atoms. With Ludwig Boltzmann as her charismatic teacher, Lise quickly confirmed that physics was her calling.

Professor Boltzmann maintained that phenomena that could not be seen directly nonetheless did exist, a viewpoint that went against the prevailing ideas of the time. In an era when science was seen as a handmaiden for industry or war, scientists like Lise Meitner, Ludwig Boltzmann, Albert Einstein and Max Planck saw scientific research as an idealistic pursuit. These theoretical physicists pushed the boundaries beyond the visible in order to advance knowledge, by justifying or disproving their theories through rigorous experimentation. Professor Boltzmann was one of the strongest advocates for the existence of atoms, something that was still disputed in the early twentieth century. Years later, Lise's nephew, the physicist Otto Frisch, said that 'Boltzmann gave her the vision of physics as a battle for ultimate truth, a vision she never lost.'

By the summer of 1905, Lise had finished her coursework and embarked on her doctoral work. At the time, in Austrian and German universities, PhD research only took a few months. Certainly this was the case with Lise. Her serious outlook on life, with no time for what she saw as frivolous activities, meant she was largely focused on work. Lise didn't marry or have children and, as far as her personal papers indicate, never had a significant romance, but she led a full life, was a devoted friend and surrounded herself with, as she described, 'great and loveable personalities' who provided a 'magic musical accompaniment'.

In February 1906, Lise was awarded her PhD in physics, with the highest honours. Her thesis on heat conduction in solids was published by the Vienna Physics Institute. Later that year, Lise was drawn to the hot topic of radioactivity and set out on what would become the predominant path of her research career. The first steps were not easy. Lise was only the second woman to obtain a physics PhD at the University of Vienna and there were no prospects for women scientists in Austria. Lise even wrote to Marie Curie, but Marie replied politely that she had no position available at that time, so Lise resigned herself to following her father's backup plan and become a school teacher.

Lise started teaching in a school in the daytime but in the evening she would go into the university's physics department to continue her research. The discovery of natural radioactivity by Antoine Henri Becquerel and the Curies in 1896 had opened the door to the study of phenomena that could not be seen directly with the naked eye. We now know that an unstable radioactive nucleus tries to become stable by releasing one of the three types of radiation they discovered, which are named after the first three letters of the Greek alphabet: alpha radiation, a helium nucleus with low penetrance; beta radiation (an electron ejected by a radioactive element) with medium penetrance; and gamma radiation, a high energy and highly penetrating ray, rather than a subatomic particle.

Working with Stefan Meyer, an assistant of Professor Boltzmann's, Lise designed experiments with alpha particles, which she used to bombard various elements, showing that scattering was greater with elements that had a larger atomic mass. This important finding would, several years later, lead Ernest Rutherford to discover the atomic nucleus.

Fortified by her successes in the laboratory, Lise felt that she had what it took to become a scientist. She would later say, 'This gave me the courage to ask my parents to go to Berlin for a few terms.' Her request coincided with a visit from Max Planck, one of the giants of theoretical physics, who, with her parents' blessing, inspired her to further her studies. At the time, Germany was considered the scientific centre of the world. The country had invested heavily in universities and technical colleges, and was home to Albert Einstein and Max von Laue and Max Planck, each of whom would later win a Nobel Prize in Physics. Arriving in September 1907, aged twenty-eight, Lise may have only intended to study for a few semesters but stayed for more than thirty years.

Max Planck became her friend and mentor, often treating her as his surrogate daughter. As German universities still had their doors firmly shut to women, she had to ask special permission to attend his lectures. Lise was the only female student that Professor Planck allowed in, an honour of which she was fully aware.

Shortly after Lise arrived in Berlin, she met Otto Hahn. Lise was small, slight and shy but she soon found a friend and collaborator in the form of Otto, who was the same age as her. Sturdy and tall, Otto had an informal, charming manner and his outgoing personality compensated for Lise's shyness. Lise remarked that she 'had a feeling that I would have no hesitation in asking him all I needed to know'. Although their friendship was later tested to the limits, they shared an innate curiosity and love for science and would remain in contact for the rest of their lives.

Otto Hahn was an expert in the chemistry of radioactive elements such as uranium and together they would make names for themselves in the field of radioactivity. The two scientists' complementary skills led to a thirty-year collaboration: Lise, trained in physics, was a brilliant mathematician who thought conceptually and could design highly original experiments to test her ideas; Otto, trained in chemistry, excelled at meticulous laboratory work.

Joining Otto to work in Emil Fischer's Institute of Chemistry in Berlin was not a smooth process. Initially, women weren't permitted, allegedly because its director was convinced they would set fire to their hair in the laboratories. Lise was given a room in the basement but forbidden to come upstairs even to talk to Otto, and had to pay a visit to a hotel down the street if she wanted to use the bathroom.

But her determination and sharp mind soon earnt her a reputation and the respect of her colleagues, even those men who had chosen to ignore her every time she walked down the corridor.

Over the next few years, Lise and Otto studied beta radiation (one of the three types of radioactive decay, which Marie Curie was also investigating) and published a number of articles in 1908 and 1909 describing the energy levels of the particle emitted in beta decay – the electron. It was a productive time as they developed both their understanding of radioactivity and their working relationship. Lise found Otto to be a congenial workmate, particularly once she discovered he shared her love of music. She said, 'Although he doesn't play an instrument, he is markedly gifted musically, with a very good musical ear and an extraordinarily good musical memory ... we would frequently sing Brahms duets, particularly when the work was going well.' It was a formal friendship, however, and for many years they didn't even share a meal together. Remarkably, the work Lise did with Otto Hahn was unpaid, yet as long as she was working in a laboratory finding a husband to support her was unlikely. Instead, her father continued to provide an allowance, as her sole source of income.

In 1912, the Kaiser Wilhelm Institute for Chemistry was established in Berlin, and Otto and eventually Lise both obtained a position there. Five years after arriving in Berlin, Lise became Prussia's first female research assistant under Max Plank. She was thrilled: 'Not only did this give me the chance to work under such a wonderful man and eminent scientist as Planck, it was also the entrance to my scientific career. It was the passport to scientific activity in the eyes of most scientists and a great help in overcoming many current prejudices against academic women.' The position also gave Lise her first salary, at the age of thirty-four.

When the First World War broke out, Otto, like most other German scientists, was drafted into the military. Lise's physics background meant she served as a nurse handling X-ray equipment. She worked gruelling hours, sometimes doing twenty-hour shifts, and witnessed the horrors of war. It was a rude awakening. In October 1915, Lise wrote to Otto, 'You can hardly imagine my way of life. That there is physics, that I used to work in physics, and that I will again, seems as much out of reach as if it had never happened and never will again in the future.' But it did and, after the war ended, Lise and Otto returned to their investigation of what they assumed would be a link between two radioactive elements, actinium (atomic number 89) and uranium (atomic number 92). In 1918, they published their findings, having discovered an element they named protactinium (atomic number 91). Their experiments established that actinium is a product of the radioactive decay of protactinium.

In recognition of this, Lise, now nearly forty, was made head of a new physics department at the Kaiser Wilhelm Institute. On 7 August 1922, she gave her habilitation lecture. Habilitation refers to the teaching qualification that is a necessary step to a professorship in many European countries. Her lecture was entitled 'The Significance of Radioactivity for Cosmic Processes' but the title was misreported as being a discussion of 'Cosmetic Processes'.

In 1926, aged forty-seven, this formerly shy young woman became an assertive professor who her nephew Otto Frisch would describe as 'short, dark and bossy'. The years between 1926 and 1938 were among the most productive and happiest for Lise. She was so besotted with physics that her work banished the insecurity of her youth and enabled her to pursue her passion. Lise went on, independent of Otto Hahn, into nuclear physics, an emerging field in which she was a pioneer, publishing fifty-six papers on her own between 1921 and 1934.

Throughout the early twentieth century, physicists' understanding of the atomic nucleus came on in leaps and bounds. Atoms are the basic units of all chemical elements. In 1911, Ernest Rutherford conducted an experiment in which he fired negatively charged electrons (discovered in 1897 by Joseph Thomson) through a sheet of gold foil. Most of the electrons passed straight through, but a few were deflected, or knocked out of alignment. From these results, Ernest Rutherford identified that most of the mass of a gold atom, and hence other elements' atoms, was concentrated in a central nucleus surrounded by electrons. Niels Bohr, and subsequently others, refined this model by demonstrating that the cloud of light-weight electrons revolves around the nucleus in a complex and erratic orbit.

In 1917, Earnest Rutherford discovered the positively charged particle in the nucleus and named it as a proton in 1920. As the decade wore on, physicists began to realise that it was the number of protons in an atom's nucleus that determined which element it was and its position in the periodic table (see page 71). The atomic number of the first element hydrogen is 1 (one proton in its nucleus), the next element helium is 2 (two protons) and so on.

During the early 1930s, Lise's research was broadening to include many complex areas of physics, from cosmic rays (atomic nuclei from outer space) to the nature of radioactivity, and she was now among Germany's foremost nuclear scientists. She had the equipment, resources and staff to turn her hand to all of these problems, often stepping in quickly to investigate new phenomena.

Lise ran a very tidy laboratory, as she had understood early on that a cleaner laboratory meant more reliable results. She would hang rolls of toilet paper next to telephones and door handles so they could be wiped clean, and handshaking was discouraged. Her drive to keep work spaces free of contaminating radioactive elements meant that, unlike the Curies, Lise and Otto Hahn did not suffer any known side effects. Both lived long and healthy lives.

Before Marie Curie began to study radioactive elements, scientists believed that atoms were impossible to divide or alter; an atom of one element could not change into another. The discovery of radioactivity demonstrated that atoms can change and that radioactive atoms release energy in the process. The discovery of radioactivity and radium in 1898 opened the door to modern physics.

In 1931, in a review article, Lise discussed how radioactive heavy elements decayed into lighter ones, and bombardment of lighter elements could be used to produce heavier ones. Neither Lise nor her colleagues foresaw a third process, the splitting of a heavy nucleus to produce lighter nuclei and the release of huge amounts of energy. While great strides were made in understanding the atom at that time, the potential significance of atomic power was not really considered. Albert Einstein said, 'loosening the energy of the atom was fruitless', and Ernest Rutherford thought, 'Anyone who expects a source of power from the transformation of atoms is talking moonshine.'

In 1932, the English physicist James Chadwick discovered the neutron – a neutral particle in the atomic nucleus. With the discovery of the neutron, there was a flurry of experiments, first of all to confirm its existence, secondly to establish that it was really an elementary particle rather than a combination of a proton and an electron, and thirdly to start investigating how it interacted with matter.

Scientists soon realised that the neutron could be used to probe the nucleus of an atom. By bombarding heavy nuclei with these neutral particles, could elements heavier than uranium (then the heaviest element in the periodic table) be created? A scientific race began between Ernest Rutherford in the UK, Irène Curie in France, Enrico Fermi in Italy and the Otto Hahn/Lise Meitner team in Berlin. The pursuit of scientific truth drove their research; none suspected where the research would lead.

Then, everything changed. In January 1933, Adolf Hitler, leader of the National Socialists or Nazi Party, was appointed Reich chancellor, head of the German state. Swiftly, he transformed Germany from a democracy to a dictatorship. In April 1933, the Nazis expelled

Jews from all places of power and influence, including those holding academic positions, such as Lise's nephew Otto Frisch and many eminent figures. Somehow, precariously, particularly as she refused to deny her Jewish heritage, Lise managed to retain an academic post for a further five years. Many of her colleagues did little to fight the growing excesses of the Nazi regime and, in 1946, Lise herself acknowledged that her decision to stay in Germany after 1933 had been 'very wrong, not only from a practical point of view, but also morally. Unfortunately this did not become clear to me until after I had left Germany.'

Lise continued to work with Otto Hahn, who was now director of the Kaiser Wilhelm Institute. In January 1934, Irène Curie (Marie's daughter) and her husband Frédéric Joliot found that bombarding light elements with alpha particles created radioactive isotopes. All the atoms of a given element have the same number of protons and electrons. The number of neutrons, however, can vary. Atoms of the same element that have different numbers of neutrons are called isotopes of that element. Some of these isotopes are radioactive.

Although it was clear that alpha particles could induce nuclear reactions, the Italian Enrico Fermi felt that the neutrally charged neutron might be more effective. Because they had no electrical charge, neutrons would be far more likely to penetrate a target nucleus than alpha particles, which had to overcome a strong repulsion from the positively charged nucleus.

Starting with the lightest element, hydrogen, Fermi and his team stared bombarding different elements with neutrons. The first to demonstrate any induced radioactivity was fluorine (atomic number 9) and others followed, so that by May 1934 Fermi had worked his way all the way up the periodic table to uranium.

There, he encountered some puzzling results. When he bombarded uranium with neutrons there were several products whose chemistry appeared different from that observed when uranium decays naturally. Uranium, which has an atomic number of 92, is naturally radioactive, although it has an extremely long half-life of nearly 4500 million years.

None of the products resembled any of the elements near to uranium in the periodic table, all the way down to radon, which has an atomic number of 86, so Fermi postulated that they had created a transuranic element, the first demonstration of an element heavier than uranium. Physicists were competing to create new elements that did not exist in the natural world. They theorised that if a neutron could insert itself into a uranium nucleus, then the atom would give off an electron and the neutron would become a proton, creating a new element with an atomic number of 93 or higher.

Lise wrote, 'I found these experiments so fascinating that as soon as they appeared . . . I talked to Otto Hahn about resuming our direct collaboration, after an interruption of several years, in order to resolve these problems. Fermi's investigations were of consuming interest to me, and it was at the same time clear to me that one could not get ahead in this field with physics alone. The help of an outstanding chemist like Otto was needed to get results.'

Along with Otto and the chemist Fritz Strassmann, Lise began bombarding uranium and other elements with neutrons and identifying the series of decay products. Otto carried out the careful chemical analysis and Lise, the physicist, tried to explain the nuclear processes involved.

Most scientists thought that hitting a large nucleus with a neutron could only induce small changes in the number of neutrons or protons in the nucleus. But, one German chemist, Ida Noddack, pointed out in September 1934 that Fermi hadn't ruled out the possibility that the uranium might actually have broken up into lighter elements. But Ida offered no further explanation and her research faded into obscurity, until Lise's interest was sparked.

Just as the research was reaching a critical point, and Lise was absorbed in the most challenging research project of her career, she found she could no longer ignore the political situation in Germany. When Hitler annexed Austria on 12 March 1938, being an Austrian of Jewish descent in Berlin was no longer merely anomalous, but perilous. The *Anschluss* (annexation) effectively made Austria part of Germany and overnight Lise lost the protection that she had enjoyed as an Austrian citizen.

Otto Hahn, who had been Lise's closest colleague for thirty years, capitulated to the pressure and told her that she must leave his institute. Lise recorded bitterly in her diary that, 'he has, in essence thrown me out'. Keen to save himself and his institute, Otto was prepared to 'sacrifice' Lise.

Although out of a job, Lise learnt that she was to be prevented on the highest authority from leaving the country. The Nazis did not want 'Jews' like Lise leaving; they wanted to imprison them and make an example of them. Scientists like Nils Bohr in Copenhagen wrote, asking her to give a lecture, a thinly veiled attempt to provide a legitimate reason for leaving Germany and enable her to seek asylum. Still, Lise resisted. Her research was her whole life, and she tried to hang on to her position as long as possible, but when it became clear she would be in danger, her friends and colleagues realised that they had to get her out fast.

The USA was an option – Lise's sister Lola was now there – but Lise felt it was too far away and too foreign. Since the *Anschluss*, as an Austrian her passport was now invalid. The Netherlands and Sweden were thought to be more lenient in this situation, and Bohr's institute in Denmark appealed as it had a reputation for excellence. On 28 June 1938, one of Niels Bohr's close colleagues in Copenhagen, Ebbe Rasmussen, came to Berlin with news. A position had been found for her in Stockholm, in the newly created institute of Manne Siegbahn, who had won the 1924 Nobel Prize in Physics for his work on X-ray spectroscopy.

On 12 July, Lise arrived early for work. She wrote in her diary later, 'So as not to arouse suspicion, I spent the last day of my life in Germany in the institute until 8 at night, correcting a paper to be published by a young associate. Then I had exactly 1½ hours to pack a few necessary things into two small suitcases.'

Her travel companion, and rescuer, was a colleague called Dirk Coster, who had been helping German refugee scientists in the Netherlands. Lise left Germany forever with only 10 marks in her purse, but before her departure, Otto Hahn gave her something to bribe the frontier guards – a diamond ring he had inherited from his mother. In the end it wasn't required and was later proudly worn by Ulla Frisch, her nephew's wife.

Lise had permission from the Dutch authorities to enter the country but, with only an obsolete Austrian passport, it was a tense moment when guards boarded the train to check travel documentation. Lise later recalled, 'I got so frightened. My heart almost stopped beating. I knew that the Nazis had just declared open season on Jews; that the hunt was on. For ten minutes I sat there and waited, ten minutes that seemed like so many hours. Then one of the Nazi officials returned and handed me back the passport without a word.'

Minutes later, she was safely across the Dutch border and, back in Berlin. Otto Hahn subsequently received a cryptic message, saying that the baby had arrived. The news came not a minute too soon, as Kurt Hess, a chemist who was an avid Nazi, had informed the authorities that Lise was about to flee.

From the Netherlands, Lise made her way to Sweden and, in the summer of 1938, she began work at Manne Siegbahn's Nobel Institute for Physics. At almost sixty years old, she was starting all over again and it was not the fairy-tale ending for which Lise had hoped. Manne Siegbahn was prejudiced against women in science and she was not made welcome. She was given laboratory space but no equipment or technical support, and not even her own set of keys. But Lise kept up her correspondence with Otto Hahn, advising him about their joint research. That way, it soon became clear that her flight from Germany had cost Hitler's regime dearly.

Otto Hahn and Fritz Strassmann were continuing with the chemical analysis of the products of the neutron bombardment experiments that they and Lise had been conducting prior to her flight from Berlin, but without Lise's expertise they had difficulty interpreting what they saw. In November 1938, Lise and Otto met secretly in Copenhagen to discuss some perplexing results.

After bombarding the nucleus of a uranium atom (atomic number 92) with a single neutron, they found what appeared to be isotopes of barium (atomic number 56) among the decay products. They were amazed to find that a tiny neutron moving at low speed would destabilise and shatter something as robust as an atom. Knocking down its atomic number and altering its chemical behaviour seemed as mythic as David taking out Goliath with a slingshot.

Radioactive transmutations were thought to happen only a little at a time. Radioactive decay would turn one element into another with very similar mass. Yet barium had barely half the mass of uranium. What was going on? Without Lise's input as a physicist, the two chemists in Berlin were not thinking along the right lines. They were focused on atomic mass – the total number of proteins and neutrons in a nucleus – rather than considering the atomic number – the number of protons.

Otto wrote to Lise in Stockholm: 'Perhaps you can suggest some fantastic explanation. We understand that it really can't break up into Barium, so try to think of some other possibility.' Lise asked Otto to repeat his experiments because she couldn't believe the results. As Fritz Strassmann would later say, 'Fortunately Lise Meitner's opinion and judgement carried so much weight with us in Berlin that the necessary control experiments were immediately undertaken.'

In late 1938, Lise had a visit from her nephew, Otto Frisch. Also an exile from Germany, he was now working as a physicist in

Copenhagen at Niels Bohr's institute. Over the Christmas holiday, Lise and Otto went for an enlightening walk. He was on skis and she was keeping up on foot, very effectively as her nephew later recounted; taking a breather, they sat on a tree trunk to mull over Otto Hahn's peculiar results. They took out a piece of paper and started calculating, and it was Lise who worked out the mathematics of exactly what it was that had happened.

Lise suggested that they view the nucleus as a water drop, following a model that had been proposed earlier by the Russian physicist George Gamow and then further promoted by Niels Bohr. As a wobbly, unstable water drop, the uranium nucleus was ready to divide itself at the slightest provocation, such as the impact of a neutron. Otto Frisch drew diagrams showing how this impact might cause the water droplet to elongate. Then, as it became pinched in the middle, it would finally split into two parts. With its resemblance to the process of binary fission in biology, where a cell splits into two, the splitting of the atom also became known as fission.

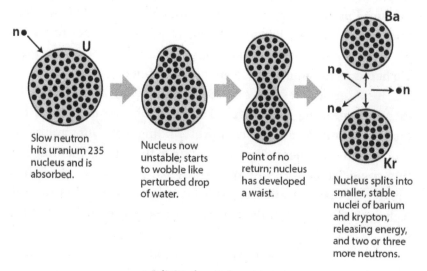

Slow neutron hits uranium 235 nucleus and is absorbed.

Nucleus now unstable; starts to wobble like perturbed drop of water.

Point of no return; nucleus has developed a waist.

Nucleus splits into smaller, stable nuclei of barium and krypton, releasing energy, and two or three more neutrons.

Splitting the uranium atom

In the face of prevailing wisdom, Lise realised what Otto Hahn had not: the uranium nuclei really were splitting in half. When overloaded by the extra neutron, the droplet – the nucleus – split in two, raising the prospect of an enormous release of nuclear energy. Not only that, the two atoms formed were of two different elements – a radical proposition in itself. The nucleus could split into several different pairs; barium (56 protons) with krypton (36 protons) or any other pair of mid-sized atoms whose protons would total 92, thus explaining the varying results of the different research teams involved.

Fission is a form of nuclear transmutation because the resulting fragments are not the same element as the original atom. The uranium nuclei could split to form barium and krypton, accompanied by the ejection of several neutrons. When the two drops separated, they would be driven apart by their mutual electrical repulsion and would acquire a very large amount of energy, about 200 MeV (Megaelectron volts). But where did that energy come from?

Fortunately, Lise remembered how to compute the masses of the resulting nuclei, realising that the total mass of the two fragments produced in the fission would be less than the mass of the uranium nucleus. From Albert Einstein's famous equation, $E = mc^2$, this missing mass would be converted into energy. As Otto Frisch concluded, 'So here was the source of the energy. It all fitted!'

The empirical results were Otto Hahn's but it was Lise who extracted meaning from them. The proof of fission required her insight as much as the chemical findings of Otto and Fritz Strassmann. No one had appreciated up to then that something like this could happen to an atomic nucleus. It hadn't crossed their minds and that's why the results were so bewildering; they were all feeling in the dark. But the part Lise played was written out of the story.

Otto Hahn's discussions with Lise took place in secret; he couldn't admit he'd been in contact with a 'non-Aryan', so Otto took her ground-breaking insight and ran with it, publishing the discovery in January 1939. He omitted her name and focused solely on the chemical process. Before publishing themselves, Lise and Otto Frisch repeated the experiment and detected both of the two fragments of the split uranium atom. They published two papers in *Nature* in February 1939, coining the term nuclear fission and explaining the physics behind the phenomenon.

Most likely Otto Hahn fought shy of incensing the Nazi authorities, although a more personal jealousy may also have played a part. Regardless, Lise felt deeply betrayed by the injustice. She wrote to her brother Walter: 'I have no self- confidence . . . Hahn has just published absolutely wonderful things based on our work together . . . much as these results make me happy for Hahn, both personally and scientifically, many people here must think I contributed nothing to it – and now I am so discouraged.' What Lise could never have anticipated was that, long after the war had ended, Otto Hahn would continue to exclude her from his version of events, a betrayal that was only tacitly acknowledged towards the end of her life.

At the time of her discovery of fission, Lise was still a stranger in a foreign country, an outsider in a male-dominated profession, living hand-to-mouth in a tiny cramped room and devastated to see her thirty years of work in Berlin being expunged from the record by Otto Hahn. Although Lise would stay in Sweden another twenty years, her enforced exile from both Germany and her rightful position at the heart of European physics meant that she never fully re-established herself in physics again.

Although scientists in Britain and the USA met the news of the discovery of fission with interest, at that time they were barely considering its use as a weapon and continued to openly publish relevant new results in the area. Three Hungarian physicists realised that, during the winter of 1939–1940, Germany was actively pursuing a nuclear weapon. They drafted a letter to the US president, Franklin D. Roosevelt, to suggest he started a programme to build an atomic bomb before the Germans did and, to add gravitas to their letter, they persuaded Einstein to sign it. Not even Roosevelt was going to ignore a letter signed by Einstein and his appeal led to the start of the Manhattan Project in 1942. By the end of 1940, the cloak of secrecy had descended and all mention of fission research had disappeared from American scientific journals.

As the war escalated, Lise continued to work, in isolation and unhappy, at Manne Siegbahn's institute in Sweden. She was desperately worried about the welfare of her German and Austrian friends, so much so that her health deteriorated and her weight dropped to less than 41kg (90lbs). In 1942, she was invited to join a group of British scientists who were going to work on the Manhattan project in Los Alamos but Lise declined. She was repulsed by the idea of a military use for fission, saying to her nephew Frisch, 'I will have nothing to do with a bomb!'

Seven years after its discovery, the process of fission was triggered inside a bomb called 'Little boy', dropped over the Japanese city of Hiroshima on 6 August 1945. The explosion wiped out 90 percent of the city and immediately killed 80,000 people; tens of thousands more would later die of radiation exposure. The bombs on Hiroshima and Nagasaki were credited with bringing World War II to an end but many historians argue that they also ignited the Cold War.

Much to her dismay, it was Lise's insight that began the nuclear age. She was deeply shocked when she heard the news of the atomic bombs and spent much of the next few days avoiding the press and trying to correct their more obvious mistakes. Lise told them nothing about the details of the bomb, mainly because she didn't know any. In one article, she is pictured 'discussing atomic fission' with a local woman in peasant dress. Lise, like the rest of the world, saw the bomb as a grave turning point for humanity. Years later, she would mourn the end of an era when, 'one could love one's work and not always be tormented by the fear of the ghastly and malevolent things that people might do with beautiful scientific findings.'

Lise was more interested in controlling the release of nuclear energy. Letting it out all at once created an unimaginable explosion; learning how to release that energy gradually would lead to a source of almost inexhaustible power. So, Lise continued her research on nuclear reactions and contributed to the construction of Sweden's first nuclear reactor.

After the war, Lise was faced with renegotiating a relationship with those of her colleagues that had stayed behind and worked under Hitler. While acknowledging her own moral failing in staying in Germany from 1933 to 1938, Lise was bitterly critical of Otto Hahn. She wrote to him, 'You all worked for Nazi Germany. And you tried to offer only a passive resistance. Certainly to buy off your conscience you helped here and there a persecuted person, but millions of innocent human beings were allowed to be murdered without any kind of protest being uttered . . .'

Clearly, Otto Hahn was very worried about his own position and had gradually dissociated himself from her – and failed to credit her work – to protect his own interests. This no doubt aided her erasure from the collective memory. Lise was nominated a number of times for the Nobel Prize but remained in scientific exile and never received it. A few weeks after the discovery of nuclear fission was made, Otto Hahn had claimed the Nobel for chemistry alone. He made a sharp separation between physics and chemistry and with the Nobel Prize structure the two are separated and the two prizes are awarded by different committees. In the 1990s, the deliberations of the Nobel Prize committee became freely available. They confirm that the set-up at the time was ill-suited to assess interdisciplinary work and Lise's omission from the 1944 award arose from a mixture of disciplinary bias, political obtuseness, ignorance and haste.

Once Otto Hahn had dictated that the discovery of nuclear fission was a chemical discovery, then it was very easy to marginalise Lise.

He suppressed and denied the value of nearly everything she had done. It was self-deception brought on by fear.

With time her role was acknowledged, although perhaps not always in quite the way Lise would have liked. On a visit to the USA in 1946, she received celebrity treatment from the American press, as someone who had 'left Germany with the bomb in my purse'. After this six-month trip to see family, combined with a visiting professorship in Washington, Lise returned to Germany. A combination of leaving Manne Siegbahn's institute and her stay in the USA helped to shake off some of the isolation she had felt for decades.

In 1947, Lise was awarded the city of Vienna's Prize for Science and Art. This was the first of an increasing number of awards in the ensuing years, culminating in the US Atomic Energy Commission's Enrico Fermi Prize, jointly awarded in 1966 to Lise Meitner, Otto Hahn and Fritz Strassmann for 'their independent and collaborative contributions' to the discovery of fission. She was the first non-American to be awarded this prize and the first woman. Here was the evidence that the international scientific community had finally recognised Lise's work on fission.

Unlike Einstein, Lise believed that after the war it was correct and proper to move on, although she only ever felt like an honoured guest in Germany, a visitor from another time. In 1947, a position was created for her at Sweden's Royal Institute of Technology. Finally, Lise had a laboratory of her own with equipment, assistants and a reliable salary. Here she worked on Sweden's first nuclear reactor, which was activated in 1954. Lise became a Swedish citizen in 1949 and continued to travel widely for work and family visits until she retired in 1953, aged seventy-five. She never considered Germany her home again. Despite the regular awards, in the years after the war it became the accepted history that Lise had nothing to with the discovery of fission. In the mid-1950s a new building on the site where she had worked for twenty-five years was dedicated to Otto Hahn alone.

Except for a few brief statements, Lise did not campaign on her own behalf. She wrote no autobiography and did not authorise a biography in her lifetime. She infrequently spoke of her struggle for education and acceptance, although the insecurity and isolation of her early career affected her deeply. Lise preferred that the essentials of her life were gleaned from her scientific publications.

Historians are still arguing about why Germany never developed an atomic bomb during the Second World War. There were scientists at the Max Planck Institute (formerly the Kaiser Wilhelm) working on it, but the bombing raids on major cities like Berlin made it very hard, and ultimately almost impossible, to carry out science there. The physicists may not have fully understood the physics, which Lise had worked out, or perhaps they never convinced the Nazi authorities to give them enough money. All the same, the Germans imagined they knew far more than the scientists in the UK and USA and were incredulous when, in August 1945, they heard of the atomic bombing of Hiroshima. In fact, the German theoretical physicist Werner Heisenberg was sure that the announcement was propaganda but, as those working on the German programme read the details, they realised that the Americans had succeeded on a scale which they had never attempted or felt possible.

Although Lise had refused to play any part in the Manhattan Project in the USA, rather to her dismay, she found herself celebrated in post-war America as the 'Jewish mother of the bomb', wrong on all counts because she never counted herself as Jewish by faith, no matter what the Nazis said. Nuclear weapons could be considered the legacy of Lise and Otto Hahn, but most of today's weapons don't use nuclear fission but the more energetic process of nuclear fusion. These are the thermonuclear hydrogen bombs developed during the Cold War. Collectively, the world's current nuclear arsenal could obliterate all life on the planet many times over.

Lise was an inspiration to physicists in their work but has also became an inspiration to those interested in nuclear non-proliferation, and nuclear disarmament issues. She felt that making nuclear weapons was an aberration and that there were more vital and beneficial applications of the technology in medicine and energy production. Her work is a prime example of how fundamental scientific research, in her case nuclear physics, can turn out to have tremendous social and political consequences.

Lise didn't predict what her science would lead to and certainly didn't want to draw attention to herself. She was ultimately concerned about the pursuit of scientific truth, and not international recognition or prizes, but would no doubt be touched that her work and its importance are increasingly being recognised. In 1982, one of the heaviest elements in the universe, element 109 in the periodic table (see page 71), was discovered and subsequently named after her – meitnerium (Mt).

Lise had always had the strength of character to speak out, both in her concerns about the applications of nuclear fission and the stance of scientists in the war years. She had long since acknowledged her failures to acknowledge the horrors of the Third Reich, but it took many years for Otto Hahn to admit any culpability. In 1958, he wrote to Lise on her eightieth birthday saying, 'We all knew that injustice was taking place but we didn't want to see it. We deceived ourselves. Come the year 1933, I followed a flag that we should have torn down immediately. I did not do so and, now, I must bear responsibility for it.' He thanked Lise for trying to make them understand and for guiding them with remarkable tact. Despite all that had happened between them, they had remained in contact and on friendly terms, a testament to Lise's ability to forgive.

In 1960, Lise retired to Cambridge, England, to be near Otto Frisch, her nephew and dedicated collaborator, who was now a professor and a fellow of Trinity College. She died peacefully in her sleep on 27 October 1968, days before her ninetieth birthday and only a few months after Otto Hahn. She was buried close to her beloved younger brother Walter, who died in 1964. Most fittingly, her gravestone in Hampshire bears an epitaph chosen by Otto: 'A physicist who never lost her humanity.'

CHAPTER NINE

Elsie Widdowson
(1906–2000)

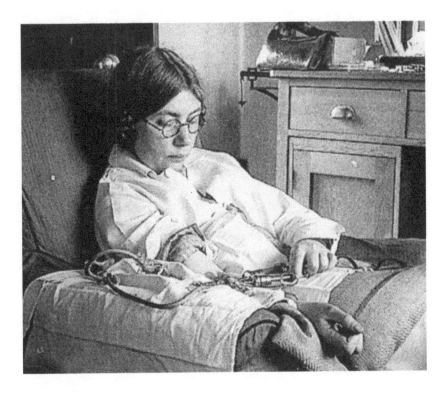

Elsie Widdowson was one of the leading British nutritionists of the twentieth century. A pioneer of the science of nutrition, she relished human experimentation, including on herself. Working with Robert McCance, she was involved in designing and implementing the Second World War ration diet, acknowledged as the healthiest diet the British population has ever had. Their work on the science behind the fortification of food earned them the title 'creators of the modern loaf'. Robert McCance and Elsie Widdowson's *Composition of Foods* remains the seminal work in nutrition and their evidence-based approach to the study of nutrition has guided nutritional advice since the 1930s.

Elsie's practical, probing, yet warm and empathic nature was instrumental to the success of her and Robert's sixty-year working partnership. Their results, linking the chemical composition of food with human health, led to the suggestion that diet in pregnancy and early life was critical for the subsequent long-term health of that child. The new field of 'developmental origins of health and disease' has refocused scientists and doctors' interest on the relevance of Elsie and Robert's work.

Elsie Widdowson was born on 21 October 1906 in south-east London to Thomas and Rose Widdowson. Her father was a grocer's assistant and she had one sister Eva, who was three years younger. Although their parents were not academics, both girls were supported in their academic endeavours and became leading experts in their chosen fields.

Elsie cycled every day to her school, the Sydenham County Grammar School for Girls. Zoology was her favourite subject in sixth form and she set out to study this at degree level. Her chemistry teacher, who Elsie admired as excellent and inspirational, persuaded her to read chemistry instead. Traditionally her school had encouraged pupils to go to one of London University's female colleges, usually Bedford College, but, encouraged by several girls in the year ahead of her, Elsie decided to go to Imperial College. As Elsie commented, it was an unusual move, 'This was a man's world with three women in our year of about 100.'

In 1924, Imperial College had growing numbers of female students and the rector took steps to improve conditions for them. Members of the Imperial College Women Students' Association were mainly the wives of professors and their president was the chemist Martha Whiteley, known by the students as the 'Queen Bee'. She helped improve the common-room environment, eventually resulting in a new common room in 1930 for both men and women. Gestures to a more feminine environment were 'some suitable vases and statuettes' and what was viewed as appropriate reading material: *Vogue*, *Theatre World* and the *Saturday Evening Post*.

After two years studying, Elsie took her BSc exams but, as was common at the time, had to spend another year at the college before her chemistry degree was awarded. Elsie's sister was now also set on an academic career path. Eva won a scholarship to read mathematics at King's College London, followed by an MSc in quantum mechanics and a PhD in nuclear physics from London University in 1938. After lecturing in physics at Sheffield University, Eva abruptly changed direction from quantum mechanics to research on bees, becoming a world authority on apiculture. Her interest in bees began when she and her husband, James Crane, received a beehive as a wedding present; in an interesting link with her sister's later work, the present giver had hoped that it would help supplement their meagre wartime sugar ration.

In her last year at Imperial College, Elsie spent time in the small biochemistry laboratory run by Professor Sammy Schryver. There she separated amino acids (the building blocks of proteins) from various plant and animal materials. Before the routine use of chromatography (an extraction technique that relies on the differential movement of components in a solution), the extractions were on a vast scale, involving bucketfuls, rather than beakers, of starting material.

After she was awarded her degree in 1928, Elsie began to look for a job. Her forays into the world of food and nutrition began with an offer from Helen Porter (née Archbold), in the Department of Plant Physiology at Imperial College, to work on the chemistry and physiology of apples. Every other week, Elsie took the train to the orchards of Kent, the 'garden of England', to harvest apples from specified apple trees. Back at the laboratory, she developed a method for separating and measuring the sugars (simple carbohydrates) in the fruit at different stages of the apples' life cycle, from ripening to storage.

Helen Archbold was instrumental in putting Elsie on her research path. As her PhD supervisor, she was always on hand to help and to advise and, in particular, taught Elsie the importance of communicating her science. Encouraged by Helen, Elsie published her findings in the *Biochemical Journal* and subsequently used this work as part of her PhD thesis.

In 1931, after the apple study was finished and she had been awarded her PhD, Elsie decided she would rather work on animals and humans than plants. For a year she learnt some human biochemistry with Professor Edward Dodds at the Courtauld Institute at Middlesex hospital in London, publishing an important paper on kidney metabolism. Then, in 1933, aged twenty-seven, she began to look for a research post, not an easy task at the time. As Elsie said, 'I didn't really want to be a dietitian but jobs in research were hard to come by for beginners in the early 1930s.' Dietetics, or the practical application of the science of nutrition, was just starting to attract interest among scientists and the medical profession. Following the lead of Professor Dodds, who saw the potential for this new area, Elsie enrolled for the first postgraduate diploma in dietetics, a one-year course, at King's College, London, in the Department of Household and Social Science.

Before starting on the course, Elsie went to work in the kitchen at King's College Hospital to learn about catering on a large scale. There, she had her first encounter with Dr Robert McCance, who was cooking joints of meat in order to learn more about their composition. Robert had originally trained in biochemistry and then qualified as a doctor in 1929. He was interested in treating diabetic patients. Eight years previously, with the discovery of insulin, it had become clear that the control of sugar levels by insulin in diabetic treatments would require detailed knowledge of the carbohydrate or sugar content of food such as fruit and vegetables.

As they chatted, Elsie realised that they shared a common interest in these sugars. Robert invited her to his laboratory where they compared notes on their previous findings, at which point Elsie plucked up enough courage to point out that his figures for carbohydrates in fruit were too low, explaining that some of the fructose sugar must have been destroyed by acid hydrolysis. Robert was more impressed than offended and went on to get a grant for Elsie from the Medical Research Council and they continued their studies together, detailing the composition of raw food, such as fruits, vegetables, nuts, meat and fish, and the changes that occurred when food was cooked.

The early work with Robert McCance and the knowledge gained from the dietetics diploma whetted Elsie's appetite for the field. And so began a sixty-year partnership, covering wide-ranging topics in nutrition, including the composition of foods, mineral metabolism, wartime diets, fortification of bread, body composition and the importance of diet in development, to name but a few.

As part of her diploma in dietetics, in 1934 Elsie was seconded to the diet kitchen at St Bartholomew's Hospital (Barts) kitchen to work with Margery Abrahams. Margery had trained in the USA where dietetics had started and she became both Elsie's colleague and a good friend. Several years later, in 1937, they wrote a book together called *Modern Dietary Treatment*.

The position at Barts was supposed to last six months but Elsie only worked there for six weeks: that was more than enough time for her to realise that the data on the composition of British foods was somewhat lacking. American food tables were being used to calculate patients' diets but these were based solely on raw food, and the carbohydrate content was calculated simply from what was left after water, protein and fat had been deducted. Importantly, this was giving inaccurate readings for diabetic patients who needed precise calculations on the sugar content of foods.

And so began a lengthy and detailed process to establish a practical set of tables showing the composition of British foods. Elsie and Robert McCance performed tens of thousands of meticulous analyses, carried out with little technical support and at a time when laboratory methods required bulk samples. Six years later, in 1940, the first edition of *McCance and Widdowson's Composition of Foods* was published, with fifteen thousand values – the first thorough analysis of the nutritional content of raw and cooked foods. Updated in numerous editions since then, it remains an authoritative and comprehensive account of nutrient data for both popular and less commonly consumed foods in the UK.

While the food composition studies were ongoing, Elsie also became involved in numerous other projects. Nutrition, as a science, was in its infancy and there was much to discover. Over the period 1934–7, Elsie and Robert investigated salt deficiency in humans, For example, Robert had noticed that diabetics, particularly those in a coma, had many metabolism problems and their urine was deficient in chloride. Chloride is an electrolyte, a mineral element in the blood and other body fluids that carries an electrical charge. It is negatively charged and works with other electrolytes, such as potassium, sodium and bicarbonate, to regulate the amount of water in the body, blood acidity (pH), muscle function and other important processes. Most of the chloride in the body comes from dietary salt and is subsequently absorbed in digestion, with any excess excreted in the urine.

Robert set out to create the same salt-deficient conditions in healthy volunteers – no mean feat, as this involved persuading his young volunteers to eat a salt-free diet and to lie, sweating profusely, in a hot-air bath for two hours a day for two weeks. After washing each volunteer down with distilled water, the washings were analysed, the volunteers' water loss was measured by weight loss, and a battery of physiological tests were performed, including on their kidney function. The results from these salt-deficiency experiments helped doctors understand the important role of fluids, and maintaining the correct chemical balance, in patients with diabetic comas, heart problems and kidney diseases.

Elsie and Robert were always enthusiastic about their science and thought nothing of experimenting on themselves. As Elsie later said, 'We did not believe that we should use human subjects in experiments that involved any pain, hardship or danger, unless we had made the same experiments on ourselves.' Their early experiments in nutritional analyses highlight the willingness of Elsie and Robert, and their equally enthusiastic volunteers, to be involved in often quite severe levels of challenge to the human body's endurance and innate regulatory mechanisms.

Many of these experiments focused on how the body handles the absorption and excretion of minerals. In the early 1930s, Robert McCance had been given some beds for his patients in King's College Hospital, where one particular patient had a lasting influence on his future research. She had polycythaemia, a condition in which there is an abnormally high number of red blood cells. After treatment with acetyl phenylhydrazine, these red blood cells were broken down and their iron content liberated. To Robert's surprise, the iron was not excreted, even when he and his colleagues repeated the experiment and injected themselves with iron. Their experiments corrected the accepted thinking that the amount of iron in the body was regulated by excretion; rather, the regulation of iron levels was achieved by absorption.

On the basis of the resultant *Lancet* paper on iron metabolism, in 1938 Robert was offered the position of reader of medicine in Cambridge, funded by the Medical Research Council. He accepted on condition that he could take Elsie with him, thereby allowing their productive partnership to continue. One of their most eventful self-experimentation episodes occurred in their first year there, which Robert later referred to as a 'slight accident'.

It involved strontium, a silvery metal found naturally as a non-radioactive element and present in tiny amounts in the human body – a trace element. About 99 per cent of the strontium in the human body is concentrated in the bones. Strontium has been used medicinally for over a hundred years; nowadays, strontium is used to treat osteoporosis (brittle bones) and in a radioactive form for prostate cancer and advanced bone cancer. Strontium chloride hexahydrate is added to toothpaste to reduce pain in sensitive teeth.

Elsie and Robert were interested in how trace elements such as strontium are absorbed and excreted by the human body, thereby allowing safe and efficient use as a therapy. Elsie designed an experiment to test the excretion of strontium, whereby she and Robert took turns to inject strontium into each other's veins every day for a week, followed by measurements of the levels in their urine and stools. Estimating the correct dose was a hit-and-miss process: the first injection of strontium lactate into Elsie's arm wasn't detected so they doubled the dose and carried on with their daily injection schedule.

By day five they had run out of their first strontium preparation and hastily had to sterilise another. On day six, lulled into a false sense of security by the apparent lack of side effects, they had lapsed into doing the experiment alone. No witnesses were in attendance when they both began to suffer from severe headaches and pronounced teeth-chattering. Fortunately, a colleague happened to visit and found the pair stricken with raging fevers. The attacks were not life-threatening and they made a full recovery, even managing to collect the necessary samples for analysis. These subsequently revealed that their second strontium preparation was impure: it was contaminated with bacterial endotoxins called pyrogens – substances that cause a fever.

Having worked through the periodic table in their excretion studies, it was a salutary lesson in the perils of self-experimentation. They were, however, pleased with the results, which showed that the human body is slow to dispose of strontium and that most of this excretion is through the kidney, not the bowel. After the Second World War, when atomic fallout was a concern, those involved were surprised that Elsie and Robert had studied the excretion of strontium in the late 1930s.

As Elsie and Robert built up their body of knowledge of the composition of foods, they turned their attention to the calculation of energy requirements and nutrient intake by individual men, women and children. Up until the 1930s, these calculations had been based on estimations of the supposed energy needs of whole families, the so-called 'man-value'. There was a pressing need to take individual needs into account, as it became clear that 'man-values' underestimated the nutritional needs of women and children and overestimated everyone's protein requirements.

Elsie initially surveyed sixty-three men and sixty-three women, publishing the results in 1936, followed by 1000 children aged one to eighteen years, obtaining detailed information on their diet. The information from these surveys revealed the wide range in intake of energy and nutrients between different individuals, even if they were the same age and sex.

In 1939, the Second World War started, and Elsie and Robert, armed with their knowledge of food composition and individual energy requirements, turned their attention to the pressing need for experiments in rationing. Like Dorothy Hodgkin (see Chapter 5), Elsie was one of Churchill's scientists. Winston Churchill (British Prime Minister 1940–5 and 1951–5) was committed to developing British science, understanding how important it was for both the war effort and for the nation's future. The little-known story of Churchill's fascination with science led to some key scientific achievements that helped Britain win the Second World War, ranging from Elsie's wartime diet to the invention of radar, the production of penicillin and antibiotics, and the top-secret research behind the first atomic bomb.

Food supplies, nutrition and rationing were of prime importance in 1939/40. Less than a third of the food available in Britain at the start of the war was produced at home but enemy ships were targeting incoming Allied merchant vessels, preventing vital supplies – including fruit, sugar, cereals and meat – from reaching the UK. The British government needed to know whether food produced solely in Britain could meet the needs and physical demands of the population. Elsie and Robert once again became human guinea pigs and, along with their colleagues, put themselves on a ration diet to see what were the minimum amounts and types of nutrients necessary to remain healthy and fit. The proposed ration diet consisted of tiny quantities by today's standards and even then was initially roundly criticised as being inadequate. Each person's diet included the following per week: 4 oz (113 g) of fat, 5 oz (142 g) of sugar, 6 oz (170 g) of home-grown fruit, one egg, 4 oz (113 g) of cheese and 16 oz (454 g) in total of meat and fish. Wholemeal bread and vegetables, including potatoes, were unrationed.

After three months on their experimental diet, which largely consisted of bread, cabbage and potatoes, they felt fit and healthy, so Elsie and Robert set off for the Lake District to test their fitness. Here they took long fell walks and bike rides along the local roads; on one day McCance cycled thirty-six miles, presumably on a bike with far fewer gears than today's bicycles, and achieved this over a change in altitude of 7000 feet, burning 4,700 calories in the process. They didn't set out to break any records, and didn't, but their fitness and stamina on the ration diet was more than adequate for their physical challenges. They concluded that the proposed wartime rations, with extremely limited dairy products, would be sufficient for the nation's dietary needs with one important adaptation.

Elsie and Robert's experiments on the ration diet had revealed that wholemeal bread interfered to a certain extent with the absorption of calcium. This was particularly important as the wartime ration diet was low in naturally calcium-rich food, like milk and cheese. They found that wholemeal bread contained a phosphorous compound, phytic acid, which formed an insoluble salt with calcium in the small intestine, thus preventing its absorption. The addition of chalk (calcium carbonate) to the bread flour improved the absorption of calcium and also provided another source of calcium.

In deference to the pure-food enthusiasts, 100 per cent wholemeal flour did not have calcium carbonate added. Different types of bread are made from flour of varying extractions. The higher the extraction, the more bran, germ and tougher outer layers of the wheat grain are in the flour. A wholemeal flour is by definition 100 per cent extraction, since it contains all the parts of the wheat grain. White flour, which is approximately 70 per cent extraction, and 70 to 85 per cent extractions of wholemeal flour, were fortified, based on Elsie and Robert's findings. This situation continues to this day in the UK, despite the massive changes in diet since the war and the sufficiency of calcium-containing dairy products. White wheat flour (not wholemeal) is still fortified with calcium carbonate, alongside iron and vitamins B1 and B3; in the UK there is a proposal to join the USA in adding folic acid to help pregnant women prevent neural tube defects, such as spina bifida, in their developing babies.

Fortification of food was an important part of Elsie and Robert's work in the Second World War. In Ireland, there was a shortage of wheat and 100 per cent unfortified wholemeal flour was used to make bread. This was associated with a rising incidence in rickets, where bones are soft and weak, sometimes resulting in malformed limbs and breakages, and usually caused by a lack of calcium, vitamin D or phosphate. Elsie and Robert were called to Dublin to describe their fortification studies to doctors and politicians, including the Taoiseach (prime minister) Éamon de Valera. Subsequently, calcium phosphate was added to a less pure flour preparation and the incidence in rickets slowly decreased.

When the war ended, the problems arising from malnutrition in parts of Europe started to emerge. Elsie and Robert's fortified bread, now containing B vitamins, iron and chalk, had attracted much interest and Elsie met with Sir Edward Mellanby to discuss the post-war loaf. He was a physician and pharmacologist, and conducted early studies into the causes of rickets, suggesting in 1919 that it was a disease of malnutrition. He told Elsie, 'There must be a lot of hungry children in Germany. You go and find out the truth about all this.' In the spring of 1946, Elsie and Robert, funded by the Medical Research Council, went to Germany. Elsie originally intended to stay for six months but this became a three-year exercise.

In January 1947, Elsie drove through the snowy countryside looking for a suitable orphanage to study. She eventually found one in Duisburg where the children, aged 5–14 years, were underweight and below average height. She carefully designed a diet that contained just 8 g of animal proteins a day and relied on bread to provide 75 per cent of the children's energy intake. The children were each given bread made from one of five flours, ranging from 100 per cent extraction (wholemeal) to 72 per cent extraction (white) with or without B vitamins and iron. All of the bread was fortified with calcium carbonate. After eighteen months, all of the children had gained height and weight rapidly and were physically greatly improved. When she brought one child, a girl, from each group to Cambridge for the British Medical Association annual conference, the audience couldn't tell which girl had been fed which bread. For all measureable growth and health purposes, the bread types were equivalent. Needless to say, the girls loved this great adventure.

While she and Robert were in Germany, Elsie visited and set up nutritional studies in several orphanages. The levels of extraction of the bread flour hadn't made a difference, but Elsie noticed some anomalies. Even with extra rations in the form of unlimited bread and margarine and jam, the children in one children's home were not growing as quickly as a comparable home where there were no extra rations given. It later transpired, after careful investigation by Elsie's research nurse, that the woman who was running the former home was unkind and singularly uncaring towards these children. In spite of the extra food, they were faring less well than their counterparts. Elsie later wrote, 'Tender loving care of children and careful handling of animals may make all the difference to the successful outcome of a carefully planned experiment.'

Elsie was one of the scientists consulted on the necessary diet to rehabilitate concentration camp victims who had suffered gross starvation. The research programmes she and Robert set up in Germany lasted for forty years and continued their work on the effects of nutrition on growing and adult humans.

In January 1949, aged forty-three, Elsie left Germany and returned to the body composition studies she had started four years earlier. Building on her analytical experience from her food composition studies, Elsie began looking at the composition of the human body at different stages of development. She used bodies dissected by the pathologist at Addenbrooke's Hospital in Cambridge and backed up her research with comparative studies on animals including pigs, rats, cats and guinea pigs. For example, the human baby contains a high proportion of body fat at birth: 16 per cent of its body weight is fat, compared to most species where it is 1–2 per cent. There were exceptions: Elsie had found a dead newborn grey seal on a beach in Scotland and bought it back to Cambridge to analyse, where she found that, like guinea pigs, its body fat was around 10 per cent.

Elsie and her colleagues such as John Dickerson discovered that mammals differ widely in their degree of development at birth and this is reflected in the composition of their organs and tissues. The composition of any one sample body as revealed by chemical analysis was like 'a snapshot of a busy street full of pedestrians and automobiles ... essentially the static reproduction of a dynamic scene, full of individual and collective activity,' according to Elsie and John Dickerson. Malnourished animals and humans were a particular focus of Elsie's, and her many studies in this area revealed the importance of sufficient nutrients for brain, skeletal and muscular development, for example.

Elsie often used farmyard animals in her nutrition experiments. Robert McCance kept pigs at his home in Cambridgeshire and Elsie worked with these pigs for over fifteen years. She showed that if you undernourished them until they were three years old, when pigs generally stop growing, and then allowed them to eat what they wanted, the pigs never caught up with their well-fed littermates. They were, however, able to produce normal-sized piglets that grew to their full adult potential.

She observed similar results with varying litter sizes in rats. In small litters where there was more milk available per rat, those that were small at weaning remained small, even if there was limitless food available after weaning. Her use of large and small litter rats subsequently became internationally famous and the combination has been used in a variety of studies worldwide. Manipulation of litter size affects milk composition, energy, protein and fat intake, intake of bioactive compounds present in milk, maternal behaviour and interaction with pups, learning behaviour and brain weight.

Another study in pigs revealed the importance of the first exposure to food. When newborn pigs were removed from the sow and given only water for the first twenty-four hours of life, this affected the development of their digestive tract. The pigs that were allowed to suckle, as normal, had a much longer and heavier digestive tract than their water-fed littermates, with a more developed local immune response in the gut. This was subsequently shown, as Elsie suggested, to be related to the absorption of antibodies and growth factors from the colostrum, the first milk, just after birth.

Extrapolating from animal experiments to humans is difficult: there are numerous confounding influences, such as heterogeneous genetic factors, diet, sex and physical activity. But Elsie's experiments, and those of other scientists, have resulted in a growing realisation of the importance of maternal and infant diets on the subsequent health and wellbeing of a developing baby.

In 1968, at the age of sixty-two, Elsie became head of the Infant Nutrition Research Division at the Medical Research Council's Dunn Nutrition Laboratory. There she became involved in the analysis of adipose tissue – the loose connective tissue, composed of adipocytes, which contains fat used for insulation and as an energy source. This new area of work came about because she had been involved in the planning of a session on infant feeding at a conference.

At the time, Dutch babies who were not breast fed were drinking a formula where all the cow's milk fat had been replaced with maize oil. The fatty acids in this were 60 per cent linoleic acid (a polyunsaturated fatty acid), compared with 1 per cent in cows' milk fat and 8 per cent in breast milk. Elsie found that these differences had a profound effect on body composition. The weight of the Dutch formula-fed babies was 10 per cent linoleic acid; this represented ten times as much linoleic acid in their bodies compared to breast-fed babies and over forty times that of the British cow's milk-based formula-fed babies.

The implications of this for future development were, and still are, unclear but Elsie started to probe these findings further in an animal model. She used guinea pigs as, like humans, the foetus lays down fat in its adipose tissues before birth. At birth, the body fat of the young of mothers fed on different oils reflected the human findings and similarly the fatty acid composition of the red blood cells, muscle, liver and the brain. Most significant was the effect on the fatty acid composition of the myelin in the brain.

Myelination occurs in utero and for some time after birth, and is one of the fundamental events in the normal development of the central nervous system. Myelin helps in the quick and accurate transmission of electrical signals carrying information from one nerve cell to another (see page 180); it protects and insulates the nerve fibres using fatty acids. The role of different fatty acids (derived from breast or formula milk) in this process is still the topic of hot debate but Elsie's results highlighted that the use of particular lipids in human nutrition must be carefully screened for possible effects on brain development.

In 1986, the year she turned eighty, Elsie travelled to Washington, DC, spending several weeks working in the Nutrition Laboratory at the National Zoo, with her friend and colleague Olav Oftedal. He had been collecting information about the milk composition of as many species of animal as possible. In 1984, Olav had been on an expedition to the pack-ice off Labrador in northern Canada to study seals. One of the species, the hooded seal, doubles its birth weight in four days on a rich diet of 6 per cent fat milk, enabling the mother to leave it and go back to sea. Olav had collected newborn and suck-led seals (killed in accordance with the Canadian Sealing Regulations) from different times of the year, including when the mothers were 'hibernating', and two years later they were waiting in the freezer ready for analysis.

With Elsie's interest in milk, and its effect on adipose tissue and the composition of animals, Olav knew that she was the person to help him with this task and they set about dissecting the animal, weighing, measuring and analysing the various parts of the body. Elsie later commented that she 'thoroughly enjoyed being associated with it all, and getting my hands, or rather rubber gloves, dirty again'. Like all her work, this resulted in the publication of several papers, revealing the composition of the rich milk that enabled the seal pups to survive for long periods untended.

In 1988, Elsie finally retired from her nutritional research but continued to keep active in other ways well into her eighties. For a number of years, she was the chair, or played a key role, on several national and international committees, adding to her previous roles as president of the Nutrition Society (1977–80), the Neonatal Society (1978–81) and the British Nutrition Foundation (1986–96). In the 1980s, Elsie received several awards in recognition of her status as a lecturer in the field, including two international lecturers' awards (1985 and 1988) and one from the Nutrition Society of New Zealand (1989).

She relished the opportunity to pass on the knowledge and experience she had gained and enjoyed finding out about people, particularly children and students, thus making sure she shared her science with them in an understandable and relevant manner. On one visit to a boarding school, she intrigued the sixth form by showing them what their predecessors had eaten in the Second World War – up to 2 oz (approximately 55 g) of meat a day. This was notable not for its small quantities – equivalent to the rations of the time – but because the school was by now strictly vegetarian.

Elsie was a pioneer in the study of nutrition. As she pointed out, 'Nutrition as a subject did not exist when I started. I have been a chemist, biochemist, plant physiologist, medical researcher and a physiologist.' The universal and enduring use of *McCance and Widdowson's Composition of Foods*, and the incorporation of their names into the title, means that they are inextricably linked with the field of nutrition. Elsie's scientific research continues to be referenced by nutritionists worldwide. Any literature search related to nutrition yields a citation bearing her name. Elsie's work shaped wartime rationing, created the wartime loaf and paved the way for later work on the damage that poor nutrition in utero and early childhood does to adult health.

Although food was scarce in the Second World War, the rations by and large produced a healthy diet, the long-term effects of which continue to emerge. Food during the war was focused on sustainability, minimising waste and nutrition. The British population has never been healthier. The low sugar, but high vegetable, diet of the ration years is now known to be hugely beneficial in reducing the incidence of diseases associated with obesity, such as type II diabetes and heart disease. The wartime ration diet may have affected intelligence, too. A study published in 2014 suggested that, even taking into account improved hygiene and medical care, children born in 1936 and living for their early childhood on wartime rations were on average over fifteen points higher on an IQ scale than those born in 1921.

The early 1940s also saw the introduction of the first nutritional standards for school meals, the first food-labelling laws, and mandatory fortification with vitamins A and D. Elsie and Robert's wartime study of calcium availability in breads led to the statutory fortification of bread with chalk, a huge contribution to the health of the nation during difficult times. This fortification of UK bread flour with calcium (combined with iron, niacin and thiamine) continues after a 2013 consultation by the Department for Environment, Food and Rural Affairs: the majority of respondents felt that the cost was low and the health benefits significant.

The National Food Survey was set up in 1940 to monitor Britain's diet and, in 1947, Elsie's Medical Research Council special report on children's diets was the first national survey of its kind; similar surveys continue to yield information and recommendations. The most recent Diet and Nutrition Survey of Infants and Young Children was published in 2011. It provided detailed information on the food consumption, nutrient intakes and nutritional status of infants and young children aged four to eighteen months living in private households in the UK. The ten or so recommendations from this include the suggestions that: babies should be exclusively breastfed for the first six months, breastfeeding mothers should take vitamin D supplements, and a range of complementary foods should be introduced to babies from six months.

Margaret Ashwell, Elsie's biographer and member of the EarlyNutrition Programming Project, said, 'She was very interested in early nutrition and what happened next . . . She knew that a baby was what their mother ate in pregnancy and how what the child ate in early life could influence their later health.' EarlyNutrition, which is funded by the European Union, has researchers from thirty-five institutions worldwide who, since 2012, have joined forces to study how early nutrition programming and lifestyle factors impact the rates of obesity and related disorders.

This ambitious programme sets out to answer such questions as: Which dietary influences increase long-term obesity risks, are there sensitive time periods for early programming? Are there specific metabolic or epigenetic (non-genetic environmental influences on gene expression) markers involved and what impact does our micro-biome, our resident bacteria and other microorganisms, have? To achieve their objectives, researchers are conducting animal studies, observational studies in large concurrent populations, and randomised controlled human intervention trials. The thorough-ness and extent of this project mirrors many of Elsie's pioneering efforts in nutritional research and she would have recognised the opportunities projects such as EarlyNutrition offer, helping to deter-mine how diet and lifestyle in pregnancy can be improved to protect the health of future generations.

In the course of her sixty-year partnership with Robert McCance, Elsie published a huge number of papers: her first in 1931 on a method for detecting reducing sugars and the last when she was eighty-six in 1992, on the birth and early development of infant physiology. The total number of papers runs to over six hundred, of which Margaret Ashwell highlighted a hundred in her biography. These illustrated the wide range of areas that Robert and Elsie had researched together, and the friendships they had made, while some papers were chosen because they had a story attached to them: one was written on board a ship bound for the USA, braving a force 9 gale in the Atlantic. Others show not only the more infor-mal style of scientific paper writing in the 1930s and 1940s but also reveal her Christian faith, peppered as they are with quotations from the Bible. Elsie delighted in pointing out that the first published account of a nutrition study (Daniel 1: 11–16) recognised the need for a control group.

Examples of Elsie's influence and impact are many and varied. In the UK, a new Medical Research Council Unit for Human Nutrition Research was set up in 1998 and named the Elsie Widdowson Laboratory. Similarly, the Food Standards Agency was established in London in 2000 and the library was named after Elsie. A fellowship award also bears her name: the Elsie Widdowson Fellowship Award at Imperial College, London (her alma mater), allows female academic staff, returning to work after maternity leave, to concentrate fully on their research work, alleviating the requirement to teach for one year.

Elsie was honoured with a number of awards in her lifetime. In 1976 she was made a fellow of the Royal Society and three years later she became a CBE (Commander of the Order of the British Empire). During the last years of her life, Elsie became the most highly honoured scientist in the UK. In 1993, aged eighty-seven, she was made a Companion of Honour. Awarded by the Queen, this British honour was founded in 1917 by her grandfather King George V. Membership of the Order is limited to sixty-five men and women who have excelled in their chosen field. Ever modest, Elsie was bemused by the honour but nonetheless delighted to receive it.

In discussing Elsie's work, it's impossible to ignore the influence of, and collaboration with, Robert McCance. He described his first encounter with Elsie in the King's College kitchens in the 1930s as a 'momentous meeting'. Their phenomenally successful partnership is an interesting illustration of the balance required between different skills and mindsets in any successful scientific research. Robert had quite a direct, sometimes difficult, manner and Elsie was recognised as an arbitrator when his criticisms were taken personally. Their approach to the research was different, too. Hamad Elneil, a Sudanese colleague, put it in a nutshell, 'The professor provided the breadth for a project and Elsie provided the depth.' Another colleague, Eric Glazer, found that 'the result is a perfect blend of their talents in which the effect is far greater than a mere summation of their skills'.

Margaret Ashwell remembers her 'otherworldliness'. Elsie's enthusiasm, drive and intuition meant that she usually found the solution to a scientific problem but she achieved this with a large degree of humility and a self-deprecatory manner. After the chairman's laudatory introduction at an dietetics' conference in Washington, Elsie remarked, 'Well, you can take all that with a pinch of salt.'

Elsie's desire to experiment extended to all aspects of her life, not only her nutritional research. Margaret Ashwell once flew with her to the USA and was somewhat perplexed by her behaviour at airport security. Elsie insisted on passing through the scanners more than once, fully aware that it had bleeped after the first time. When questioned about her behaviour, she explained that she had removed her hearing aid the second time round and the interference had substantially decreased, allowing her to conclude that it was this, and not her hairpins or her suspenders, that had set the machine off.

At home, having never married, Elsie lived alone in a thatched cottage near the Cam at Barrington in Cambridgeshire, where she enjoyed growing vegetables and had a large apple orchard. These provided ample supplies for, as Margaret Ashwell describes, her many visitors, 'who would always be greeted as if they were the most important person in the world'. Margaret, who had joined the British Nutrition Foundation in 1988, lived only twelve miles from Elsie and regularly travelled to London with her by train.

When Margaret suggested writing a joint biography of Elsie and Robert McCance, Elsie was horrified, saying, 'Whoever would want to read a book about us?' Evidently, they agreed and *McCance and Widdowson: A Scientific Partnership of 60 Years* was published in 1993. Robert McCance died a year later, after a fall at home, while Elsie survived him by seven years. In 2000, she had a stroke while on holiday in Ireland and later died at Addenbrooke's Hospital in Cambridge.

Elsie's unassuming manner, combined with a razor-sharp mind and a strong vein of common sense, inspired those around her, not least the young scientists she regularly encountered. Her advice to those trying to make their way in science was practical and insightful, born out of a long career and her fair share of trials and tribulations in the field: 'If your results don't make physiological sense, think and think again! You may have made a mistake (in which case own up to it) or you may have made a discovery. Above all, treasure your exceptions. You will learn more from them than all the rest of your data.'

Elsie's wide-ranging experiments in the field of nutrition were instrumental in developing this vital area of science and led to realistic and informed recommendations on diets for everyday life.

Responding to a question about her own diet and longevity, Elise, who lived until she was ninety-four years old, replied, 'I eat butter, eggs and white bread, which some people think are bad for you but I do not. However, I do eat plenty of fruit and vegetables and drink lots of water. I think my longevity is largely due to my genes [her father died aged 96 and her mother aged 107]. Perhaps having been breast-fed has helped!'

CHAPTER TEN

Chien-Shiung Wu
(1912–97)

Chien-Shiung Wu was a Chinese-American who became one of the finest experimental physicists of her generation. Able to perform delicate and intricate experiments that many of her contemporaries found impossibly hard to do, she showed in an extremely complicated experiment in 1956 that an accepted 'law' of physics, the law of parity, was, in fact, wrong. When the Nobel Prize was awarded the following year for this important discovery, however, it went to the two theoreticians who explained it; Chien-Shiung received nothing. Although she was overlooked by the Swedish Academy, she became one of the most celebrated female physicists of the twentieth century, breaking down barriers and allowing other women to follow in her footsteps.

Chien-Shiung Wu was born on 31 May 1912 in the town of Liuhe in the Jiangsu province of China, which lies on the east coast and is home to China's historical capital, Nanjing. She was the second of three children in a time before China enforced its 'one child' policy; the two other children were boys. Chien-Shiung was extremely close to her father, who encouraged her interests, and she grew up in an environment surrounded by books, magazines and newspapers. One could say that she was destined for great things from birth as her name means 'courageous hero'. At the time that Chien-Shiung was born, girls in China were only just starting to be allowed to go to school. In the highly structured and traditional Chinese society, it had not been deemed necessary for girls to get a formal education; their main role in life was to support their future husbands. Luckily for Chien-Shiung, her father Wu Chung-i had established a girls' school, Ming De School, the first in her province and one of the first to be founded in China.

Aged eleven, after her initial education at her father's school, she left her hometown to go to the Women's Normal School No. 2 in Suzhou, a larger town about thirty miles to the west of Liuhe. This was a boarding school, and students were split between those who would train to become teachers and those who would have a regular education. Getting a place on the teacher training route was far more difficult, but, if successful, a student's tuition and board were paid for, and students were guaranteed a job upon graduation. Chien-Shiung chose the more competitive teacher-training route, and was ranked ninth among about ten thousand applicants.

Chien-Shiung finished at the school in 1929, graduating top of her class. She gained a place at the National Central University in Nanjing. She was expected to continue to study to be a teacher, which was one of the few paths opening up to educated women in China at that time. She confided to her father that her real desire was to study physics, and he went out and bought some advanced mathematics and physics books, which Chien-Shiung devoured over the summer to prepare herself for university.

Chien-Shiung initially concentrated on mathematics at university, but she later transferred to her real love, physics. During her time as an undergraduate (1930–4), she also became quite involved in student politics. At the time, Imperial Japan was trying to exert its influence in the region, and relations between China and Japan were quite tense. Japan would go on to occupy the Manchuria province of China in 1931 until the end of the Second World War. Chien-Shiung was elected to be one of the student leaders, partly because it was felt that, as one of the best students at the university, her involvement in student politics would be forgiven, or at least overlooked, by the authorities.

After graduating, Chien-Shiung became a teacher in a small university, but she felt that her physics education was far from complete. She obtained a research assistant position at the National Academy of Sciences (Academia Sinica) in Shanghai, working in the field of X-ray crystallography. This is where Chien-Shiung learnt to do experimental physics, an area in which she would go on to become a world leader. Although she enjoyed her work, she still hankered after a formal physics education, but at that time in China there were no graduate programmes in physics open to women.

Her supervisor in her job at the Academia Sinica was Professor Gu Jing-Wei, who had returned to China after obtaining his PhD at the University of Michigan in the USA. He encouraged Chien-Shiung to follow a similar path and head to the USA to get her doctorate and to return to China with the expertise and experience she would gain there. She applied to the University of Michigan and was offered a place on their graduate programme, so, with financial help from her uncle, she set off in August 1936 with a female friend, Dong Ruo-Fen. Her parents and uncle came to see her leave on the SS *President Hoover*. That was the last time she saw them.

The ship arrived in San Francisco, and soon after her arrival Chien-Shiung met the physicist Yuan Chia-Liu. He was the grandson of Yuan Shikai, the first president of the Republic of China, who had also proclaimed himself Emperor of China. Yuan took her to see Raymond Birge, the head of the physics department at the University of California, Berkeley, and he offered her a place on the graduate programme. Chien-Shiung never adopted an 'English' name, unlike so many Chinese who move to the West. She was either known as Miss Chien-Shiung or later Madame Chien-Shiung. She held on to both her Chinese name and many Chinese traditions for the rest of her life, and never adopted Western dress.

Chien-Shiung quickly showed her abilities, making rapid progress in both her formal classes and in her research. She was ostensibly supervised by the physicist Ernest Lawrence, but her day-to-day supervision fell to Emilio Segrè, an Italian-American physicist who is best known for discovering the anti-proton in 1955, for which he won the 1959 Nobel Prize. Her PhD thesis had two separate parts. The first part was a study of a phenomenon called *bremsstrahlung*, which is a German word that literally means 'breaking radiation'. It is the X-ray radiation produced by electrons when they rapidly slow down. The second part was to investigate the production of radioactive isotopes of xenon produced by the bombardment of uranium using the 37-inch and 60-inch diameter cyclotrons at the Radiation Laboratory at Berkeley.

In order to investigate the *bremsstrahlung*, Chien-Shiung used phosphorous-32, a radioactive isotope of phosphorous that emits beta particles. Beta particles are high-speed electrons and can produce X-rays when they are rapidly slowed down as they pass through matter. This was the first time that Chien-Shiung would work with beta decay, a subject on which she was to become the world authority by the late 1940s. She investigated the X-rays being produced by the beta particles slowing down in matter, and she devised experimental methods to distinguish different energies of the X-rays being produced. She and Emilio Segrè were also able to document the complete chain of radioactivity for different elements. They were able to identify every nucleus that was created from the fission of the beta-emitters that she was studying, and each element that was formed as the radioactive elements decayed.

Chien-Shiung completed her PhD in June 1940. Highly recommended by Ernest Lawrence and Emilio Segrè to the management at Berkeley, she was unable to secure a faculty position as none were open to women at the time. Instead, Chien-Shiung stayed working at the Radiation Laboratory as a post-doctoral research fellow. She travelled the USA giving talks about her research. Although she had learnt English before she came to the United States, she never quite mastered the language. At times her pronunciation and grammar rendered what she was saying unintelligible. To try to avoid misunderstanding, Chien-Shiung would write out her talks meticulously beforehand, rehearse them over and over again, and follow these notes as closely as she could during her lectures.

If Chien-Shiung was falling short of mastering English, she had no such problems mastering experimentation. In fact, she was already gaining a reputation in the field of nuclear physics for having an extraordinarily meticulous and careful eye for detail. She became adept at working out what was important in experiments, what might lead to errors and spotting other scientists' mistakes. In 1941, in a paper for *Physical Review*, she wrote, 'The large discrepancies between their experimental results and theoretical calculation may be explained by the fact that part of the external X-rays excited on the [magnetic] pole faces and walls has gotten into the counter.'

Things were moving in the right direction on the personal front, too. Yuan Chia-Liu had moved to study at the California Institute of

Technology, but in May 1942 they married at the home of Robert Millikan, one of the founders of Caltech and a Nobel Laureate in Physics. Neither Chien-Shiung nor Yuan's families were able to attend the wedding due to the cost and because the USA and Japan were at war. Upon finishing his doctorate, Yuan was offered a research position at RCA Laboratories in Princeton on the East Coast, where he worked on radar as part of the American war effort.

Chien-Shiung became a faculty member at Smith College, a prestigious all-women's college in Northampton, Massachusetts. She and Yuan would meet each weekend in New York, midway between Princeton and Northampton. However, Chien-Shiung found her job at Smith frustrating as her duties primarily involved teaching and there was hardly any opportunity to carry on with her research. In order to be with Yuan, she chose to accept an offer from Princeton University. Although she was hired to teach there, at the time women were not admitted as undergraduates, a situation she found very hard to accept.

Her primary task when she arrived was to teach naval officers. They were being sent to Princeton to further their understanding of engineering, but her physics classes were necessary as the subject lies at the foundation of engineering. Chien-Shiung became well known for her research, but she was also a committed teacher and cared about the welfare of her students. As she later said, 'They were good students, but they were afraid of physics, and first you had to get them over the fear.'

After only a few months at Princeton, Chien-Shiung went to Columbia University in New York City for an interview with Harold Urey and Gilbert Lewis, who had both previously been at Berkeley. Enrico Fermi, one of the greats of nuclear physics, had advised his colleagues that the easiest way to solve some of the technical difficulties in producing a sustained nuclear chain reaction would be to ask Chien-Shiung to help them. They were devising ways to enrich uranium in order ultimately to make an atomic bomb. Uranium occurs in two natural isotopes, U-235 and U-238. U-238 is stable, but U-235 is radioactive. For a uranium bomb, the amount of U-235 needed to be increased, as natural uranium is 99.284 per cent U-238, which is of no use in a bomb. The process of increasing the amount of U-235 became known as enrichment, and several approaches were tried to achieve this. Columbia was tasked with one of the approaches being investigated.

As it was part of the Manhattan Project, this research was classified. Urey and Lewis, working for the Division of War Research, interviewed Chien-Shiung for two gruelling days to find out if she was suitable for the job. The two were careful not to give Chien-Shiung any of the details of the research they were doing, but at the end of the two days they asked Chien-Shiung if she had any idea what it was that they were doing. She smiled and said, 'I'm sorry, but if you wanted me not to know what you're doing, you should have cleaned the blackboards.'

Chien-Shiung was hired immediately. From March 1944, she worked in what were publicly called the 'Substitute Alloy Materials Laboratories' at Columbia; no one outside of the Manhattan Project knew that the people in the laboratory were working on a project to build the world's first atomic bomb. She was asked to help enrich the uranium by taking U-238 and turning it into the fissile U-235. Using the process known as 'gaseous diffusion', Chien-Shiung was able to obtain extremely pure samples of U-235, a vital part in building the bomb. Without the successful work being done at Columbia, of which Chien-Shiung played a vital part, the USA may never have been able to develop and deploy the atomic bomb.

After the war had finished, Chien-Shiung had to make the decision as to which area of research she was going to concentrate her efforts. After careful thought, she decided to work on unravelling the details of beta decay, which was still poorly understood. This choice would define a significant part of her career and propel her to international acclaim. Columbia University asked her to stay on after her excellent work on the gaseous diffusion process. She was relieved, as she had chosen to specialise in an area, nuclear physics, in which there were virtually no jobs outside of academia. In many branches of physics, industries are keen to employ physicists, but this was not true for nuclear physicists, so Chien-Shiung's options were limited.

Chien-Shiung, because she was a woman, could not be taken on as part of the teaching faculty at Columbia but she was given a job as a research professor. As a research professor, there was no guarantee of permanency but as she was not required to do any teaching, she could concentrate on research full-time.

In 1933 Fermi, who later became the first person to achieve a sustained nuclear chain reaction, had developed a theory of beta decay. Since the publication of his theory, physicists had been trying to experimentally verify or disprove his theory, with varying results. It was an area that suited Chien-Shiung. Ever since her teenage years she had wanted to do great things, to 'transform our perception of the underlying structure of the universe'. Chien-Shiung was a very practical scientist: she did not just choose something to study based on mere curiosity – it had to serve a purpose or she felt that it was not worth doing.

She became very adept at improving equipment, including the neutron spectrometer, to allow more precise experiments to be conducted. A spectrometer is an instrument that splits up radiation or particles into different energy bins. With light this would be frequency (or wavelength), but with neutrons it would be their kinetic energy, their speed. The existing neutron spectrometers could not differentiate between the small energy differences in the neutrons being emitted from radioactive nuclei, so Chien-Shiung set about correcting this by improving the sensitivity of Columbia's neutron spectrometer. She redesigned the device to allow the spectrometer to take twice as many measurements at one time and updated the electronics so that it could respond in thousandths of a second.

Using this improved neutron spectrometer, she was able to meas-ure and characterise the neutrons being emitted by radioactive forms of cadmium, iridium and silver. Although not ground-breaking experiments, they did improve our knowledge of nuclear physics and added to Chien-Shiung's reputation as a careful and precise experimentalist. Chien-Shiung also started thinking about the distance range of nuclear forces, two of the four forces in nature.

The first force of nature to be understood was gravity, when Isaac Newton developed his law of universal gravitation in the late 1600s. In the mid-1800s James Clerk Maxwell did the same for electromag-netism, at the time the only other force of nature known. Later, with the discovery of radioactivity and the atomic nucleus, two new forces came to light: the strong nuclear force and the weak nuclear force.

Briefly speaking, the weak nuclear force is responsible for radio-active decay, and the strong nuclear force is responsible for holding the atomic nucleus together. The nucleus is made up of two types of particles, protons and neutrons. Whereas neutrons carry no electri-cal charge, protons carry a positive charge. This means that they repel each other due to the electromagnetic force; so why don't nuclei fly apart? The answer is that the strong nuclear force is an attractive force between all particles in a nucleus, whether they be neutrons or protons. We call it the strong force because it is much stronger than electromagnetism, and it is what stops protons from repelling each other, but only when they are very close.

The strong nuclear force therefore has a very short range, only a few times larger than the diameter of a proton. Such small distances are measure in fermi, named after Enrico Fermi. A fermi is 10^{-15} of a metre – one million billionth of a metre (for comparison, an atom is millions of times larger). Julian Schwinger, a theoretical physicist based at Harvard at the time, had theorised that the range of the strong nuclear force should either be 0 fermi or 8 fermi; but neither prediction seemed to be correct. In a set of delicate experiments using her improved spectrometer, Chien-Shiung was able to show that the measured value was about 3 fermi, a result that the physics community accepted as by now they knew the precision of Chien-Shiung's experimental technique.

Chien-Shiung had the long-term aim of understanding beta decay. She felt that not only would it be an important area, but that she could use her ability to perform exquisitely precise experiments to solve the problem of whether Fermi's theory of beta decay was correct or not. What was already known was that the emitted beta particles (electrons) could have an energy between 0 and 0.6 Megaelectron volts (MeV). An electron volt is an extremely small amount of energy; the usual unit of energy, a joule, is far too big for the energies at these atomic and nuclear levels, so we use the electron volt instead. Fermi's theory predicted what number of electrons should have an energy of 0.1 MeV, what number should have an energy 0.2 MeV, etc., when electrons are emitted via beta decay.

We call such a distribution of energies an energy spectrum. The energy spectrum observed in experiments, however, did not agree with that predicted in Fermi's theory. Many electrons had far less energy than the theory predicted, but there was a huge variation in the experimental results and no one could agree on the observed distribution. Chien-Shiung felt that solving this problem played to her strengths but it was not enough to do the experiment correctly: she also needed to be able to explain why others were getting differing results.

She felt that the radioactive sources that others were using in their experiments were too thick. She reasoned that many of the electrons, after being emitted via beta decay, would ricochet (scatter) from other atoms in the source as they emerged and lose energy as they did so. This, she felt, would account for the observed greater number of low-energy electrons compared to what the theory predicted, and differing source thicknesses could also explain the varying results that the various teams were getting.

Creating a thinner source was not easy. The spectrometers that were being used to do this experiment used a large iron core to produce a magnetic field, and this was an essential component of how they were able to separate particles with different energies. For the spectrometer to work, it was necessary for the source to have a small surface area but to produce a large number of beta particles. The only way that this could be achieved was to make the source thick, so that its volume was large even though its surface area was small.

To get around this problem, Chien-Shiung, following her success in redesigning the neutron spectrometer, decided to redesign the spectrometers used to measure beta decay. Instead of using the iron-core spectrometer, Chien-Shiung found an old spectrometer in the Columbia laboratory that used a coil of wire, a solenoid. This older spectrometer, once she had brought it back to operating condition and modified it, would allow her to use large area sources and hence keep them thin.

The other part of the problem was to learn how to make thin film sources. As no one had needed to do this before, Chien-Shiung and her team had to invent the process. Not only did the sources need to be thin, but the thickness needed to be uniform. She outlined her remedy in a paper she submitted to *Physical Review* in 1949, writing, 'A more uniform source can be obtained by adding a trace of detergent to the $CuSO_4$ [copper sulphate] solution'. Adding a bit of soap was the key!

When Chien-Shiung did the experiments with the modified equipment and reagents, she found excellent agreement between Fermi's theory and her results. She was able to verify that it was the thick sources that had led to the erroneous results. She had finally solved the problem of beta decay, and she gained worldwide praise from the nuclear physics community. Chien-Shiung wrote up her methods in the research literature, and soon laboratories around the world were copying her methods to produce their own thin films and were able to verify Chien-Shiung's results.

This example shows the qualities that made Chien-Shiung an experimental physicist almost without equal. While other researchers would tend to redo existing experiments, usually replicating the set-up of the experiment that they were verifying, Chien-Shiung went her own way. She designed and used new instruments to enable her to measure more accurately than anyone else. As William Fowler, a Caltech physicist who won the 1983 Nobel Prize, said, Chien-Shiung's 'beta decay work was important for its incredible precision'.

Chien-Shiung's work on beta decay was deemed by many of her peers to be worthy of a Nobel Prize. Beta decay had been studied since the late 1890s when Ernest Rutherford first realised that there were different kinds of radioactive decay, and Fermi's theory had been around for more than ten years. Physicists throughout Europe and the USA had done dozens of experiments to try to see whether Fermi's theory was correct or not, but Chien-Shiung was the first to do the experiment precisely enough to remove all the errors in others' experiments.

Yet the rules of the Nobel Prize stated that it should only be awarded for a discovery. Chien-Shiung had not discovered anything new – she had confirmed Fermi's theory – so strictly speaking her work, important as it was, was not eligible. At the time many felt that the Nobel Prize was being awarded more for politics than for scientific merit. Robert Friedman, in *The Politics of Excellence*, wrote, 'The mid-1940s Nobel Prizes were not awarded on the basis of recognising merit; instead they had become to a great extent instruments in the politics of science.' It would not be the last time that Chien-Shiung was to carry out work worthy of a Nobel Prize, only to be overlooked.

After Chien-Shiung and Yuan married in 1942, they experienced the typical life of dual academics, sometimes living together but other times being separated by their respective jobs. In 1947, they had a son, whom they named Vincent. The family lived in an apartment on Claremont Avenue, New York, two blocks from the campus of Columbia University. Chien-Shiung would live in this apartment for more than fifty years; it was sufficiently close that she could spend long days in her beloved laboratory.

In the 1940s and 1950s, once a woman was married she was expected to give up her career and look after her husband and any children, but Chien-Shiung never considered doing such a thing. She was completely devoted to her research and having a son was not going to stop her dedicating her time to her work. In fact, she had a hard time taking any time off; she was a complete workaholic. As a graduate student related, on one occasion her students arranged for Chien-Shiung and Vincent to go to a children's movie, not only to give her the evening off work, but so that they could get on with research without her constant 'tinkering'. They soon realised that the plan had failed. Chien-Shiung appeared at work in the evening, as usual: she had given her ticket to her nursemaid.

Chien-Shiung expected her students to share her drive and standards. She frowned on her students when they took time off, even once chastising a student for being away on a religious holiday. As one of her former students Noemie Koller said, 'It was very exciting, but she was rough – very demanding. She pushed the students until they did it right. Everything had to be explained to the last decimal. She was never satisfied. She wanted people to work late at night, early in the morning, all day Saturday, all day Sunday, to do things faster, to never take time off.'

After her extensive work on beta decay, Chien-Shiung decided to turn her attention to one of the most energetic processes in nature: particle/antiparticle annihilation. When matter and antimatter come together they produce a huge amount of energy, far more even than in nuclear fission. For example, if we were to bring 1 kilogram of matter and 1 kilogram of antimatter together, the energy released would be more than 2000 times more than the energy released in the atomic bomb that was dropped on Hiroshima.

It was Paul Dirac who had first predicted the existence of antimatter in 1928, and the positron (the antiparticle of the electron) was experimentally discovered in 1932 by Carl Anderson. But, since then, very little experimental work had been done to test the theory of anti-matter. Chien-Shiung felt that this was a ripe area for experimental exploration, largely unchartered. When an electron and a positron annihilate they produce two high-energy gamma rays. Theory predicted that the two gamma rays would have certain characteristics, but this had never been tested, so no one knew whether the theory was correct. The characteristics were the states of polarisation of the two photons (light particles).

Polarisation is a relatively easy concept to understand. Wearing polarising sunglasses reduces glare by only allowing light that is orientated in a particular direction to pass through. Light travels like a wave, and as James Clerk Maxwell had shown in the nineteenth century, it is in fact composed of a varying electrical field at right angles to a varying magnetic field. When we talk about the orientation of a light-wave, we are talking about the orientation of the electric field, as this is much stronger than the magnetic field.

When light comes from a source like the Sun, the orientation of the waves is random. Polarising filters, as used in polarising sunglasses, only allow light with an orientation that is vertical to pass through. As any light reflected from the surface of water has a polarisation that is horizontal, only allowing vertically orientated light through reduces reflections (and hence glare) from water surfaces.

With the two photons produced when an electron and positron annihilated, the theory predicted that they would have polarisations at right angles to each other, and that the ratio of one polarisation to the other was precisely a factor of two. This theory had never been tested successfully, so Chien-Shiung wanted to devise an experiment to do so. Others had tried to test this prediction: one group had errors which were so large that their results were essentially useless; another group had found a result that did not match the ratio of polarisation that the theory had predicted.

To perform this difficult experiment, Chien-Shiung would use the cyclotron at Columbia University and a radioactive form of copper to produce the positrons. These would be smashed into electrons, and counters and detectors would be carefully placed around the apparatus to measure the photons emitted. In an experiment that lasted thirty hours, she kept one detector fixed in the same place while she moved the other around in a great arc, measuring the emitted photons at different angles. Then she interchanged the detectors and repeated the same thirty-hour operation. The experiment showed that the ratio of the two polarisations was 2.04, very close to the 2.0 predicted by theory.

With Chien-Shiung's increasing reputation, it was perhaps inevitable that China would try to entice her back to her homeland. The National Central University of China offered her a position, and also one for her husband Yuan. On the face of it, such an offer would seem impossible to turn down, but Chien-Shiung did exactly that. The officials at the National Central University gave her a year to reconsider but Chien-Shiung realised that, if she were to return to China, she might not be allowed back into the USA. The Communist Party was in control in mainland China and relations between the USA and China were worsening. As Chien-Shiung had stayed in touch with her father, she asked him for advice. Although he desperately wanted to see her, he told Chien-Shiung that it was not a good time to return to China. She and Yuan decided to stay and became US citizens in 1954. She had been made an associate professor at Columbia two years earlier, thus giving her the security of a tenured position. She was the first female scientist to be given tenure at the university, and in 1958 she was promoted to full professor.

Another important concept in nuclear physics, in fact in physics in general, is parity. This is less familiar than the idea of polarisation, as we don't have parity glasses! If a spinning top is not spinning too fast, it is quite easy to determine in which direction it is spinning (either clockwise or anti-clockwise), but if we look at the spinning top in a mirror it will, of course, appear to spin in the opposite direction as the image is reversed. We accept that this is not a different reality: it is just that the mirror transforms the reality to look different. We describe this difference as a *parity transformation*.

A similar concept of parity exists in the physics of subatomic particles. It was introduced in 1924 by Otto Laporte, a German-born American physicist. He was describing how atoms emit light, and by assigning a parity to each state of the system, he was able to show that parity is conserved (does not change) when an atom emits a photon. Now, if parity is conserved it means that the process will look exactly the same in a mirror. This is different from the spinning top example, in which the parity is transformed rather than conserved. By 1927, this idea had been extended by Eugene Wigner to a fundamental principle of physics. He proved that all interactions involving the electromagnetic force conserve parity. It seems that down at the atomic level parity is conserved, and this idea was accepted for the next few decades as being true.

Chien-Shiung had been doing research on a radioactive form of lead, ^{210}Pb, with funding from the Atomic Energy Commission. Until 1939, it had been assumed that the radioactive decay of ^{210}Pb was well understood but, as better measurements were made by Chien-Shiung and others, the picture became more confused. No two experiments agreed. Using her customary care and attention to detail, Chien-Shiung was able to properly measure the gamma rays and electrons (beta particles) being emitted by ^{210}Pb and settle the controversy concerning its decay. Around this time, Maria Goeppert-Mayer, a German-born American physicist working at the University of Chicago, developed a model of the atomic nucleus, which became known as the shell model. In this theory she described different energy levels within the nucleus, and the theory would win her the 1963 Nobel Prize in Physics. As part of her theory, the nuclei were assigned values of intrinsic parity, just as Laporte had earlier done for atoms. In the experiments that Chien-Shiung was doing on ^{210}Pb, she had to consider the parity of the nucleus.

One of the biggest puzzles around this time was the so-called 'theta-tau problem'. The two particles had been discovered in 1949: as particle accelerators like the cyclotron became more powerful, they were creating more and more new particles. Some measurements indicated that the theta and tau particles were really one and the same, but other experiments suggested that they were two different particles. But, if they really were two different particles, it would mean that parity was not conserved in nuclear reactions, something nearly all physicists were loath to abandon.

Chien-Shiung decided that she would settle the matter once and for all. Particle physics, born from experiments with particle accelerators after the Second World War, had discovered essentially three types of particles. Electrons appeared to be fundamental particles, but when electrons were smashed into either protons or neutrons, a whole slew of new particles came streaming out. Physicists started trying to figure out what was going on and found out that protons and neutrons are examples of what they called baryons, but that other types of particles being created in these experiments had quite different properties, and these were called mesons.

The theta and tau particles mentioned above were examples of a meson. The theta meson decayed into two pions (another type of meson), while the tau meson decayed into three pions. This clearly suggested that they were two different particles. What confused the situation was that both the theta meson and the tau meson were found to have the same mass and the same lifetime before they decayed. Surely this was too much of a coincidence, hence the feeling some physicists had that they were the same particle. It was known that the theta meson decayed into a state with positive parity, but the tau meson decayed into a state with negative parity. This was the puzzle; on the one hand the two particles had identical masses and lifetimes, but on the other they decayed into particles with different parity.

On 3 April 1956, nearly two hundred physicists gathered at the Sixth Annual Rochester Conference. One of the main topics of discussion was the theta-tau problem. Basically there were two possibilities: either they were separate particles and parity was conserved, or they were the same and parity was broken. Remember, Wigner had shown that parity must be conserved in interactions involving the electromagnetic force, but, at the scale of the nucleus, it was the strong and weak nuclear forces that were at play, not the electromagnetic force. For radioactive decay it is the weak nuclear force that is responsible. There was no proof that parity was conserved for the weak nuclear force: physicists had just assumed that it was.

A few weeks after the Rochester Conference, Tsung-Dao Lee and Chen-Ning Yang, two of the leading theoretical physicists of the time, met in a café in New York City and discussed whether parity could be violated in interactions involving the weak force. After this meeting, Tsung-Dao asked Chien-Shiung whether she knew of any experiment which had shown that parity is conserved or violated in beta decay, and she stated that she did not. Tsung-Dao and Chen-Ning realised that no one had ever shown that the conservation of parity, proven to hold for electromagnetic interactions, also held for weak interactions. They wrote a paper on the matter, and it was published in *Physical Review* on 22 June 1956 as 'Question of Parity Conservation in Weak Interactions'.

When Tsung-Dao Lee had asked Chien-Shiung's advice, he had mentioned that some people had suggested using nuclei produced in nuclear reactions or from a nuclear reactor to test the conservation of parity in beta decay, but Chien-Shiung felt this would not work. As she later said in an interview, 'Somehow I had great misgivings about using either of these two approaches. I suggested that the best bet would be to use a ^{60}Co [cobalt with an atomic mass of 60] beta source.' Doing this test of parity conservation became the most important experiment of Chien-Shiung's already illustrious career. The protons in an atomic nucleus have a property called spin. It is only an analogy, but you can think of the proton as being like a tiny little planet, spinning on its axis in a particular direction. If you place an atomic nucleus in a magnetic field, the spins of the protons will be forced to all line up, and this establishes a net spin to the nucleus, rather than the different spins of the protons all cancelling each other out as they would in the absence of an applied magnetic field.

By convention, we call a clockwise spin 'up'. From the discussion about looking at a spinning top in a mirror, we know that a clockwise spinning nucleus will appear to be 'down' (anti-clockwise) if we look at it in a mirror. If parity is conserved, the emitted particles from the reaction should emerge in equal numbers in the up and down directions. If they do not, it would look different in a mirror. So, Chien-Shiung needed to count the number of electrons coming out of the ^{60}Co source with 'up' spin, and the number with 'down' spin. If the numbers were equal, parity was conserved; if they were not, equal parity was violated.

The conversation with Tsung-Dao Lee took place the same year that Chien-Shiung and Yuan were celebrating twenty years since Chien-Shiung had left China and their meeting in Berkeley. They had planned a cruise to China to celebrate. A cabin had been booked on the *Queen Elizabeth*; it would have been their first long holiday together in many years. But, when the time came, Yuan sailed alone. Chien-Shiung knew from her conversations with Tsung-Dao that he and Chen-Ning Yang were about to submit their paper on parity violation to the *Physical Review*, and Chien-Shiung felt that she could not risk being away when the paper came out: she had to be in the laboratory doing her experiment.

Chien-Shiung, in planning her experiment in her head, knew that she would have to cool the ^{60}Co in order for the spins of the protons to align properly in a magnetic field. This is because thermal energy (temperature is just a measurement of thermal energy) would cause the protons to deviate away from being aligned, even when the magnetic field was applied. She calculated that she would need to cool the ^{60}Co to about 0.01 K (one hundredth of a degree above absolute zero, or -273 °C).

Chien-Shiung Wu in her laboratory in 1963

Her laboratory in Columbia did not have the ability to cool objects to such low temperatures. In fact, at the time there were only two or three places in the whole of the USA where it could be done, and one of them was the Low Temperature Laboratories at the National Bureau of Standards (NBS) in Washington, DC. Chien-Shiung contacted the laboratory's director, Ernest Ambler, and asked him if he could help her experiment, and he agreed. She prepared thin layers of ^{60}Co by putting it on a substrate of cerium magnesium nitrate (CMN).

Although Chien-Shiung tried to be as secretive as possible about her experiment, not wanting to be beaten to it by anyone else, word did get out. Wolfgang Pauli, the theoretician who had come up with the idea of the neutrino in 1930, asked the nuclear physicist Victor Weisskopf what was going on in nuclear research in the USA. Weisskopf told Pauli that Chien-Shiung was working on an experiment to see whether parity was conserved in weak interactions. Pauli, well known for his strong views and abruptness, wrote back, 'Mrs Chien-Shiung is wasting her time. I would bet you a large sum of money that parity is conserved. I do not believe that the Lord is a weak left-hander, and I am ready to bet a very large sum that the experiments will give symmetric results.'

By the time this letter from Pauli reached Weisskopf, Chien-Shiung had already shown that parity was indeed violated. Her first attempt failed: one of the many problems she encountered in doing this incredibly difficult experiment was that the alignment of the nuclei in the thin layer of ^{60}Co was quickly destroyed. Chien-Shiung suspected that this was due to the thin layer of ^{60}Co heating up, but she did not know why. She guessed that it could be due to the thick layer of CMN substrate, so she decided that she needed this to be a single crystal. She did not know of anywhere she could obtain such large crystals so, to quote Chien-Shiung herself, 'Relying purely on ingenuity, determination and luck, three of us (an enthusiastic chemist, a dedicated student and myself) worked together uninterruptedly to grow about ten large perfect translucent CMN-single crystals by the end of three weeks. The day I carried these precious translucent crystals with me back to Washington, DC, I was the happiest and proudest person in the whole wide world.'

Two days after Christmas in 1956, Chien-Shiung was joined by Ernest Ambler and other members of the National Bureau of Standards. Some detectors counted electrons, others counted gamma rays. Ernest Ambler jotted down numbers in his notebook. The laboratory was church-quiet, except for the noise of the vacuum pumps keeping air away from the CMN crystals and the hiss of compressors from the great refrigerators keeping the ^{60}Co at fractions of a degree above absolute zero. The scientists themselves could barely breathe as they waited in anticipation to see whether one of the 'laws' of physics would be broken.

Their initial results suggested that parity was indeed broken. The excitement subsided, however, when repeated experiments did not duplicate this result. After several more days of careful testing and checking, Chien-Shiung and her small team were finally convinced that they were indeed seeing a smaller number of electrons being emitted with one direction of spin compared to the opposite direction. Parity was not conserved! Chien-Shiung had shown that a cherished law of physics was, in fact, not true.

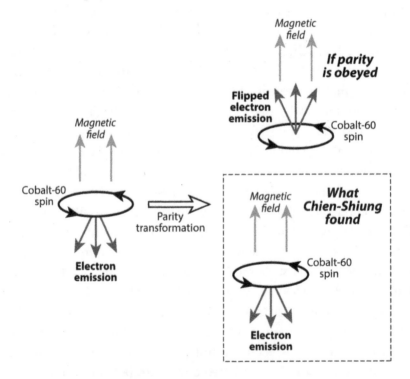

If parity was preserved, when the orientation of the experiment was reversed, the direction of the emission of electrons would change. Instead, Chien-Shiung found that the electrons always shot out the same way relative to the cobalt atoms' spin, showing that the weak force violates the law of parity.
(Source: https://galileospendulum.org/2014/03/08/
madame-wu-and-the-backward-universe)

Chien-Shiung hastily submitted two papers to *Physical Review*. A press release was also sent out, and the next day the *New York Times* carried a front-page headline, Basic Concepts in Physics Reported Upset in Tests. Not only had Chien-Shiung shown that a basic law of physics was not true, but her result also solved the theta-tau problem. If parity was violated, the theta meson and tau meson were, in fact, the same particle, which we now call the K meson. The K meson is able to decay following two different paths because parity is not conserved.

Chien-Shiung's parity-violation experiment was recognised world-wide as a masterpiece of careful experimental work. So important was this work that the Nobel Committee, which often take years to recognise the importance of a piece of work, gave the 1957 Nobel Prize for the discovery that parity was violated, but Chien-Shiung did not receive it. Instead, the prize went to Tsung-Dao Lee and Chen-Ning Yang for their theoretical work on how it may be violated. This is, undoubtedly, one of the worst oversights of the Nobel Committee. Chien-Shiung herself never complained, and by the 1960s she had started winning most of the other major awards in physics, both national and international, and was awarded several honorary degrees.

Although the demonstration of parity violation is considered Chien-Shiung's greatest success, another that vies for that title came from her work on the weak nuclear force. In the late 1940s, physicists had developed a theory called quantum electrodynamics which posited that the electromagnetic force could be explained by the exchange of particles (photons) between charges. This theory was so successful that it led to physicists thinking that all the four forces of nature (gravitational, electromagnetic, strong nuclear and weak nuclear forces), could be explained by theories involving the exchange of particles. A theory was developed that argued that the beta rays emitted by, for example, C-12 (carbon with six neutrons) and N-12 (nitrogen with five neutrons) would have the same 'shape' – the same number of beta particles emitted by each radioactive atom at each energy.

Chien-Shiung was a specialist in the weak nuclear force after her decades of work on beta decay, and felt that she could measure these energies more accurately than anyone else. She started off by redesigning (again) her spectrometer and finding ways to ensure that stray beta particles did not bounce off its walls. Others had tried to do this experiment, but their results were not conclusive, yet it did not take Chien-Shiung long to test the particle theory for the weak nuclear force. She found exact agreement between the theory and her results. She published her findings in *Physical Review Letters* in 1963 in a paper entitled 'Experimental Test of the Conserved Vector Current Theory on the Beta Spectrum of B^{12} and N^{12}'. The results were not merely a confirmation of a theory for the weak nuclear force: they provided a new way of understanding the fundamental forces of nature. We no longer think of the four forces as having fields and field lines, as Michael Faraday and James Clerk Maxwell had done in the mid-1800s. Instead, forces are understood to be due to the exchange of particles.

Towards the end of her career, Chien-Shiung decided to turn her considerable experimental talents to biophysics. She began by investigating the essential protein in red blood cells, haemoglobin. Her work looked at the detailed structure of haemoglobin, and in particular the electronic structure of the iron that carries oxygen in red blood cells. She was able to show that the iron was the same in both healthy and unhealthy red blood cells, and yet some red blood cells show a high affinity for oxygen, while unhealthy ones do not. She made considerable advances in the study of sickle-cell anaemia, bringing her expertise from a different field to bear on an area of research that, she felt, could benefit humanity a great deal. Chien-Shiung's foray into biophysics would be her last research project. As her retirement approached, she focused on promoting women in science and travelled extensively, giving talks about her life of research. When she visited her homeland, she was referred to as 'the Madame Curie of China'. She showed that it was possible, as a woman, to have a successful career in science and be a mother and wife.

In 1997, at the age of eighty-four, Chien-Shiung died in New York after suffering a stroke. She left behind a legacy that still resonates: two generations of students who learnt their craft under her guidance, important work on understanding the workings of the nucleus and one of the fundamental forces of nature, and, above all, she was an inspirational figure who showed that determination, hard work and dedication can overcome the obstacles of sex and race. One of her former graduate students, Noemie Koller, wrote, 'There has been much progress since Chien-Shiung Wu first landed in California in 1936, both in physics and in the recognition of professional women, much of it due to the perseverance, enlightenment and accomplishment of women like her. We will remember the spirit of dedication to science and to her people that characterized Chien-Shiung Wu.'

Glossary

....

Adipose tissue – loose connective tissue that stores energy in the form of fat.

Alpha decay – process in which an alpha particle is ejected from the nucleus of a radioactive atom.

Alpha particle – one of the three types of radioactivity. It consist of two protons and two neutrons bound together. Identical in structure to a helium nucleus, it is produced during alpha decay.

Amino acid – the building block of proteins. Of the twenty amino acids in the human body's proteins, nine are essential and have to come from dietary sources, as human cells cannot manufacture them.

Anaesthetic – a substance that induces insensitivity to pain.

Antagonist – a drug or a compound that opposes the physiological effects of another, for example by blocking the receptor for the agonist (a substance that fully activates its specific receptor).

Antibiotic – a medicine that can inhibit the growth of or destroy microorganisms such as bacteria.

Antibody – protein produced by specialised immune cells in the blood that fights disease by attacking, and facilitating the killing of, harmful microorganisms, such as viruses and bacteria.

Antimatter – composed of antiparticles. Each particle of *matter* has a corresponding antiparticle of *antimatter*. For example, the antiparticle of an electron is known as a positron. It has the same mass as an electron, but an opposite electric charge (positive rather than negative).

Antiparticle – a subatomic particle having the same mass as a given particle but opposite electric or magnetic properties. Every subatomic particle has a corresponding antiparticle.

Astronomy – scientific study of the universe.

Atom – basic unit of matter. It consists of a dense nucleus surrounded by negatively charged electrons circulating the nucleus in an electron cloud. The nucleus is made up of positively-charged protons and neutral neutrons.

Atomic bomb – created using the process of fission or fusion.

Atomic mass – the total number of protons and neutrons in the nucleus of a single atom.

Atomic number – the number of protons in the nucleus of an atom. This determines an element's place in the periodic table as well as its chemical properties, such as boiling point, and how it reacts to other elements.

Atomic weight – the average mass of all of the naturally occurring isotopes of an element.

Autoimmune disease – disease that results when the immune system mistakenly attacks the body's own tissues. Examples include multiple sclerosis, type 1 diabetes, rheumatoid arthritis and systemic lupus erythematosus.

Baryon – a subatomic particle, such as a nucleon or hyperon, that has a mass equal to or greater than that of a proton.

Base pair – a pair of complementary bases in a double-stranded nucleic acid molecule, consisting of a purine in one strand linked by hydrogen bonds to a pyrimidine in the other. Cytosine always pairs with guanine, and adenine with thymine (in DNA) or uracil (in RNA).

Beta decay – the process in which a beta particle is ejected from the nucleus of a radioactive atom.

Beta particle – one of the three types of radioactivity. It is an electron or positron ejected from the nucleus of a radioactive atom.

Biochemistry – the study of chemical processes in living organisms.

Botany – the scientific study of plants.

Cepheid – a variable star that has a regular cycle of brightness, with a frequency related to its luminosity, so allowing estimation of its distance from the earth.

Chemical formula – a set of chemical symbols showing the elements present in a compound and their relative proportions.

Chemotherapy – the treatment of disease with chemical compounds, especially the treatment of cancer by cytotoxic and other drugs.

Compound – substances that contain atoms of at least two elements chemically combined. Compounds are represented by formulae that show how many atoms of each element are in the compound.

Congenital – a disease or physical abnormality present from birth.

Constellation – a group of stars forming a recognisable pattern that is traditionally named after its apparent form or identified with a mythological figure.

Cyclotron – a type of particle accelerator.

DDT – dichlorodiphenyltrichloroethane is a synthetic organic compound used as an insecticide. DDT tends to persist in the environment and become concentrated in animals at the top of the food chain. DDT is now banned in many countries.

Decay series – a series of decay processes or transformations in which one element decays to create a new element, or product, that may be radioactive. The chain ends when a stable element or isotope is formed.

DNA – deoxyribonucleic acid is a self-replicating material that is present in nearly all living organisms as the main constituent of chromosomes; DNA carries genetic information in the form of genes.

Ecology – the scientific study of the relationships of organisms to other organisms and their environment.

Efficacy – the ability to produce the desired or intended result (commonly used when analysing the results of drug trials).

Electromagnetism – the phenomenon of the interaction of electric currents or fields and magnetic fields.

Electron – a subatomic particle with a negative charge. It orbits the nucleus of an atom inside an electron cloud.

Element – a substance made up of only one type of atom (with the same atomic number). Chemically elements are the simplest substances and cannot be broken down in a chemical reaction.

Elementary particle – a minute subatomic particle that cannot be divided. By definition, a particle is considered to be elementary only if there is no evidence that it is made up of smaller constituents. There are thirty-one known elementary particles including those grouped as leptons, quarks or bosons. The proton, for example, is not an elementary particle because it is made up of three quarks, whereas the electron is because it seems to have no internal structure.

Embryo – an unborn or unhatched offspring in the process of development. In humans, an embryo refers to the offspring in the period from the second to the eighth week following fertilisation, after which it is usually called a foetus.

Enzyme – a substance produced by a living organism which acts as a catalyst to bring about a specific biochemical reaction.

Epidemiology – the study of the incidence, distribution and possible control of diseases and other factors relating to health.

Fatty acid – the building block of fat and lipids.

Fauna – animals that live in a particular area or period in time.

Fission – binary fission is a process of cell division in biology. In physics, nuclear fission is a reaction in which an atom's nucleus breaks down into smaller units. Fission typically gives off neutrons and photons as well as a huge quantity of energy. Fission is a form of transmutation because the resulting fragments are new elements.

Flora – plants that grow in a particular area or period in time.

Fusion – nuclear fusion in physics is a reaction in which two nuclei combine to form a nucleus with the release of energy. It is the process that powers the Sun and the stars.

Galaxy – a system of millions or billions of stars, together with gas and dust, held together by gravitational attraction.

Gamma ray – one of the three types of radioactivity. A type of electro-magnetic radiation with extremely high energy. It occurs natu-rally during the decay of radioactive isotopes and causes damage to human tissues, including cancer.

Ganglia – a group of nerve cells forming a nerve centre, especially one located outside the brain or spinal cord. The two types are a) sensory ganglia, which receives signals from the periphery and sends them to the brain, and b) autonomic ganglia, where the signal travels in the opposite direction.

Geology – the scientific study of the history of the solid Earth, as recorded in rocks, in particular the structure of rocks and the processes by which they change over time.

Growth factor – a naturally occurring substance capable of stimulat-ing cellular growth, proliferation, healing and cellular differentia-tion; typically acts as a signalling molecule between cells.

Half-life – the amount of time required for a substance (for example, a medicinal drug or an unstable radioactive atom) to reduce or decay to half its original quantity.

Histology – the study of the microscopic structure of an organism's tissues and cells.

Immunosuppression – partial or complete suppression of the immune response.

Insecticide – chemical compound that is used to kill insects.

In vitro – outside a living organism, in a test tube or a culture dish for example (from the Latin for 'in glass').

In vivo – in a living organism (from the Latin for 'in the living').

Ion – an atom with a positive or negative electrical charge created by an uneven number of electrons and protons.

Ionisation – addition or removal of an electron to create an ion. Losing an electron creates a positive ion; gaining an electron creates a negative ion.

Isomorphous replacement – a characteristic of some molecules where substitution for one or more elements by others does not change the crystal structure; most commonly used in the determination of protein structures, where it is possible to derive isomorphous crystals of native protein and of heavy-atom derivatives.

Isotope – variation of an element. It has the same number of protons as the element but a different number of neutrons. Carbon 12, carbon 13 and carbon 14 are all isotopes of carbon: all have six protons, but with six, seven or eight neutrons respectively. Different isotopes of the same element occupy the same position in the periodic table.

Light-minute – the distance light travels in a vacuum in one minute, approximately 18 million kilometres.

Light-year – a unit of astronomical distance equivalent to the distance that light travels in one year, which is 9.4607×10^{12} km (nearly 6 million million miles).

Luminosity – related to the brightness of a star, luminosity is the total amount of energy emitted by a star, galaxy or other astronomical object per unit of time.

Magellanic Cloud – one of two irregular galactic clusters – the Large Magellanic Cloud (LMC) and the Small Magellanic Cloud (SMC) – in the southern heavens that are the nearest independent star system to the Milky Way and orbit Earth's galaxy.

Magnitude – the apparent magnitude of a star is a number that is a measure of its brightness as seen by an observer on Earth. The brighter an object appears, the lower its magnitude value.

Meson – a subatomic particle that is intermediate in mass between an electron and a proton and transmits the strong interaction that binds nucleons together in the atomic nucleus.

Metabolism – the chemical processes that occur within a living organism in order to maintain life.

Metabolite – a substance formed in, or necessary for, metabolism.

Milky Way – the galaxy of which the Sun and the solar system are a part and which contains the myriads of stars that create the light of the Milky Way.

Molecule – forms when two or more atoms (of the same or of different elements) chemically bond with each other.

Muscle tone – background tension, or level of firmness, of resting muscle tissue.

Nebula – a cloud of gas and dust in outer space.

Neonatology – subspecialty of paediatrics that consists of the medical care of newborn infants, especially the ill or premature newborn.

Neurobiology – the biology of the nervous system.

Neurology – the branch of medicine that covers the structure, function and diseases of nerves and the nervous system.

Neuron – the fundamental unit of the nervous system, a nerve cell that receives and sends electrical signals over long distances within the body.

Neutron – a subatomic particle with no electrical charge and found in the nucleus of atoms.

Nuclear chain reactions – a series of nuclear fissions (splitting of atomic nuclei), each initiated by a neutron produced in a preceding fission.

Nuclear force – the strong nuclear force is an attractive force between protons and neutrons that keep the nucleus together, and the weak nuclear force is responsible for the radioactive decay of certain nuclei.

Nuclear physics – the physics of atomic nuclei and their interactions, especially in the generation of nuclear energy.

Nucleic acid – consists of either one or two long chains of repeating units called nucleotides, which have a nitrogen base (a purine or pyrimidine) attached to a sugar phosphate molecule. The two main nucleic acids are DNA and RNA.

Obstetrics – the branch of medicine and surgery concerned with childbirth and midwifery.

Palaeontology – the scientific study of fossil animals and plants.

Parasite – a plant or animal that lives on or in a host and derives nutrients from that host while offering no benefit in return.

The relationship is sometimes a neutral one or it can harm the host.

Particle accelerator – an apparatus for accelerating subatomic particles to high velocities using electric or electromagnetic fields. The accelerated particles can be made to collide with other particles, either as a research technique or for the generation of high-energy X-rays and gamma rays.

Particle physics – the study of the properties, relationships and interactions of subatomic particles.

Period – in astronomy the orbital period of a star or planet is the time it takes to return to the same place in the orbit.

Periodic table – devised by the Russian scientist Dmitri Mendeleev in 1869, this chart arranges the elements according to their atomic numbers and chemical properties.

Pesticide – chemical compound that is used to kill pests, including insects, rodents, fungi and unwanted plants (weeds).

Pharmacology – the study of the uses, effects and actions of drugs and medicines.

Photon – the smallest discrete amount or quantum of electromagnetic radiation. It is the basic unit, the fundamental particle, of all light.

Physiology – the scientific study of how people's and animals' bodies function.

Pion – a meson with a mass approximately 270 times that of an electron.

Pitchblende – a brown to black mineral ore containing uranium and radium.

Polarisation – the process of transforming unpolarised light into polarised light. A light-wave that is vibrating in more than one plane is referred to as unpolarised light. Polarised light-waves are light-waves in which the vibrations occur in a single plane.

Positron – also known as an anti-electron, this is the antimatter to the electron. It has the same mass but the opposite electrical charge.

Protein – a large molecule composed of one or more chains of amino acids. Proteins are required for the structure, function and regulation of the body's cells, tissues and organs.

Proton – a subatomic particle with a positive charge, found in the nucleus of an atom, that cancels out the negative electron charge of the orbiting electrons. The number of protons determines the atomic number of an element and hence its position in the period table.

Purine – a two carbon-nitrogen ring base that is a component of nucleic acid. Two examples, adenine and guanine, are both found in the nucleic acids of DNA and RNA. They pair with their complementary pyrimidines.

Pyrimidine – a one carbon-nitrogen ring base that is a component of nucleic acid. Thymine and cytosine pair with their complementary purines in DNA molecules to create the double helical structure.

Radioactivity – some substances are radioactive because the nucleus of each atom is unstable and can decay, or split up, by giving out nuclear radiation in the form of alpha particles, beta particles or gamma rays.

Spectroscopy – the analysis of the interaction between matter and any portion of the electromagnetic spectrum. Traditionally, spectroscopy involved the visible spectrum of light, but X-ray, gamma and ultraviolet spectroscopy are also used.

Stellar parallax – the apparent shift of position of any nearby star (or other object) against the background of a distant object.

Subatomic particle – a component of an atom, including the proton, neutron and electron.

Supernova – a star that suddenly increases greatly in brightness because of a catastrophic explosion that ejects most of its mass.

Thalidomide – a chemical compound used as a sedative or for morning sickness in pregnant women. Withdrawn after children (of mothers who took it in pregnancy) were born in the 1960s with congenital abnormalities or missing limbs.

Theoretical physics – the description of natural phenomena in mathematical form.

Transit – in astronomy, this describes when at least one celestial body, such as a star, moves across the face of another celestial body, hiding a small part of it, as seen by an observer at a particular vantage point.

Transmutation – the change of one element or isotope to another as a result of a nuclear reaction.

Transuranic element – an element with an atomic number greater than 92 (the atomic number of uranium); such elements do not occur naturally and are created in the laboratory. They are unstable and undergo radioactive decay to form other, more stable, elements.

Uranium – metallic element with an atomic number of 92. The most common isotope of uranium is U-238, which accounts for 99.3 per cent of natural uranium. The remaining 0.7 per cent is U-235, which is less stable and can therefore be used in nuclear reactions.

Tau – an elementary particle.

Theta – an elementary particle.

Universe – all existing matter and outer space. The universe is believed to be at least ten billion light years in diameter and contains a vast number of galaxies; it has been expanding since its creation in the Big Bang about thirteen billion years ago.

Variable star – a star whose brightness changes, either irregularly or regularly.

X-ray – an electromagnetic wave of high energy and very short wavelength, which is able to pass through many materials, but not all, and leave an image on photographic paper. In medical imaging, dense areas in the human body, such as bone, allow fewer X-rays through and so appear paler on X-ray images. Less dense areas, such as skin and many tissues, allow more X-rays though. These appear black on X-ray images.

X-ray crystallography – the study of molecular structure by examining diffraction patterns made by X-rays beamed through a crystalline form of the molecules.

Zoology – the scientific study of animals.

Further Reading

Virginia Apgar

Apel, Melanie Ann, *Virginia Apgar: Innovative Female Physician and Inventor of the Apgar Score* (Rosen Central, 2004).

Apgar, Virginia and Joan Beck, *Is My Baby All Right?* (Simon & Schuster, 1972).

Rachel Carson

Carson, Rachel, *Under the Sea-Wind* (Simon & Schuster, 1941).

Carson, Rachel, *The Sea Around Us* (Oxford University Press, 1951).

Carson, Rachel, *The Edge of the Sea* (Houghton Mifflin, 1955).

Carson, Rachel, *Silent Spring* (Houghton Mifflin, 1962).

Jameson, Conor Mark, *Silent Spring Revisited* (Bloomsbury, 2012).

Souder, William, *On a Further Shore: The Life and Legacy of Rachel Carson* (Random House, 2012).

Marie Curie

Cobb, Vicki, *DK Biography: Marie Curie* (Dorling Kindersley, 2008).

Curie, Eve, *Madame Curie: A Biography* (DaCapo Press, 2001).

Goldsmith, Barbara, *Obsessive Genius: The Inner World of Marie Curie* (W. W. Norton, 2005).

Quinn, Susan, *Marie Curie: A Life* (DaCapo Press, 1996).

Gertrude Elion

MacBain, Jennifer, *Gertrude Elion: Nobel Prize Winner in Physiology and Medicine* (Rosen Central, 2004).

McGrayne, Sharon Bertsch. *Nobel Prize Women in Science* (Birch Lane Press, 1993).

St Pierre, Stephanie, *Gertrude Elion: Master Chemist* (Rourke Enterprises, 1993).

Dorothy Hodgkin

Ferry, Georgina, *Dorothy Hodgkin: A Life* (Granta, 1998).

McGrayne, Sharon Bertsch, *Nobel Prize Women in Science* (Birch Lane Press, 1993).

Thiel, Kristin, *Dorothy Hodgkin: Biochemist and Developer of Protein Crystallography* (Cavendish Square, 2017).

Henrietta Leavitt

Burleigh, Richard, *Look Up! Henrietta Leavitt, Pioneering Woman Astronomer* (Simon & Schuster, 2013).

Johnson, George, *Miss Leavitt's Stars: The Untold Story of the Woman Who Discovered How to Measure the Universe* (W. W. Norton, 2006).

Rita Levi-Montalcini

Dash, Joan, *The Triumph of Discovery. Women Scientists Who Won the Nobel Prize* (Julian Messner, 1991).

Hitchcock, Susan Tyler, *Rita Levi-Montalcini: Nobel Prize Winner* (Chelsea House, 2005).

Levi-Montalcini, Rita, *In Praise of Imperfection* (Basic Books, 1988).

McGrayne, Sharon Bertsch, *Nobel Prize Women in Science* (Birch Lane Press, 1993).

Yount, Lisa, *Rita Levi-Montalcini: Discoverer of Nerve Growth Factor* (Chelsea House, 2009).

Lise Meitner

Calvin, Scott, *Beyond Curie: Four Women in Physics and Their Remarkable Discoveries, 1903 to 1963* (IOP Concise Physics, 2017).

Conkling, Winifred, *Radioactive! How Irène Curie and Lise Meitner Revolutionized Science and Changed the World* (Algonquin Young Readers, 2018).

Rife, Patricia, *Lise Meitner and the Dawn of the Nuclear Age* (Birkhäuser, 1999).

Sime, Ruth Lewin, *Lise Meitner: A Life in Physics* (University of California Press, 1997).

Elsie Widdowson

Ashwell, Margaret, *McCance & Widdowson: A Scientific Partnership of 60 years* (British Nutrition Foundation, 1993).

Widdowson, Elsie, *All Creatures Great and Small . . . Adventures in Nutrition* (The Nutrition Society, 2006).

Chien-Shiung Wu

Calvin, Scott, *Beyond Curie: Four Women in Physics and Their Remarkable Discoveries, 1903 to 1963* (IOP Concise Physics, 2017).

Hammond, Richard, *Chien-Shiung Wu: Pioneering Nuclear Physicist* (Facts on File, 2009).

Tsai-Chien, Chiang, *Madame Wu Chien-Shiung: The First Lady of Physics Research* (World Scientific, 2013).

Index

Page numbers in *italic* refer to illustrations